II International Conference "Cognitive Robotics" 2017

IOP Conference Series: Materials Science and Engineering
Volume 363

Tomsk, Russia
22 – 25 November 2017

ISBN: 978-1-5108-6342-2
ISSN: 1757-8981

Printed from e-media with permission by:

Curran Associates, Inc.
57 Morehouse Lane
Red Hook, NY 12571

Some format issues inherent in the e-media version may also appear in this print version.

Printed by Curran Associates, Inc. (2018)

For permission requests, please contact the Institute of Physics at the address below.

Institute of Physics
Dirac House, Temple Back
Bristol BS1 6BE UK

Phone: 44 1 17 929 7481
Fax: 44 1 17 920 0979

techtracking@iop.org

Additional copies of this publication are available from:

Curran Associates, Inc.
57 Morehouse Lane
Red Hook, NY 12571 USA
Phone: 845-758-0400
Fax: 845-758-2633
Email: curran@proceedings.com
Web: www.proceedings.com

Table of contents

Volume 363

II International Conference "Cognitive Robotics"

22–25 November 2017, Tomsk, Russian Federation

Accepted papers received: 9 May 2018
Published online: 24 May 2018

Preface

Papers

1. Abstract

This issue of the journal is dedicated to the scientific and applied research presented at the II International Conference "Cognitive Robotics". The conference was held on November 22–25, 2017 in Tomsk (Russia).

The conference is organized by:
The Russian Foundation for Basic Research.
National Research Tomsk State University. Tomsk, Russia.
National Research Tomsk Polytechnic University. Tomsk, Russia.
JSC Scientific-Research Institute of Semiconductor Devices, Tomsk; Russia.
Polzunov Altai State Technical University; Barnaul, Russia.
Karaganda State Technical University, Karaganda, Kazakhstan.
Information Satellite Systems named after Academician MF Reshetnev.
Association of Innovative Enterprises and Organizations of Tomsk and Tomsk Region.
"Radio-wave imaging" JSC. Tomsk, Russia.
Anhalt University of Applied Sciences, Germany.

Sponsors: National Instruments Corporation (USA).

CONFERENCE SECTIONS: 1. Cognitive technologies and artificial intelligence in control systems for land and air-based robotic systems, and robotic submarines. 2. Sensing technology for land and air-based robotic systems, and robotic submarines. 3. Design of robotic systems (UAVs, AUVs). 4. Psycho-emotional and socio-humanitarian aspects of robotics. 5. Educational robotics

2. Committee:

Conference chairman: Vladimir I. Syryamkin, D.Eng.Sc., professor, Head of the Department of Quality Control of the Faculty of Innovative Technologies, National Research Tomsk State University; academician of International Academy of Sciences of Higher Education and Russian Academy of Natural Sciences; Director of TSU scientific and educational center «Recognition: navigation, diagnostics, mechatronics».

Conference co-chairman - Eduard V. Galazhinsky, Doctor of Science (Psychology), Professor, academician of the Russian Academy of Education, Rector of Tomsk State University, Professor.

Conference co-chairman – Stanislav V. Shidlovskiy - D.Eng.Sc., Dean of the Faculty of Innovative Technologies of National Research Tomsk State University; Head of the Laboratory of high-performance reconfigurable systems (NR TSU); Professor of Tomsk Polytechnic University.

3. The Organizing Committee of the II International Conference "Cognitive Robotics" wishes to thank the IOP Conference Series Materials Science and Engineering-Team for collaboration.

Peer review statement

All papers published in this volume of *IOP Conference Series: Materials Science and Engineering* have been peer reviewed through processes administered by the proceedings Editors. Reviews were conducted by expert referees to the professional and scientific standards expected of a proceedings journal published by IOP Publishing.

Numerical study of propagation of forest fires in the presence of fire breaks using an averaged setting

S I Marzaeva and O V Galtseva
Tomsk Polytechnic University, Tomsk, Russia
E-mail: rabota2013tpu@mail.ru

Abstract. The forest fires spread in the pine forests have been numerically simulated using a three-dimensional mathematical model. The model was integrated with respect to the vertical coordinate because horizontal sizes of forest are much greater than the heights of trees. In this paper, the assignment and theoretical investigations of the problems of crown forest fires spread pass the firebreaks were carried out. In this context, a study (mathematical modeling) of the conditions of forest fire spreading that would make it possible to obtain a detailed picture of the change in the temperature and component concentration fields with time, and determine as well as the limiting condition of fire propagation in forest with these fire breaks.

1. Introduction

The forest fires are very complicated phenomena. At present, fire services can forecast the danger rating of, or the specific weather elements relating to, forest fire. There is need to understand and predict forest fire initiation, behaviour and spread. This paper's purposes are the improvement of knowledge on the fundamental physical mechanisms that control forest fire spread. A great deal of work has been done on the theoretical problem of how forest fire spread. Crown fires are initiated by convective and radiative heat transfer from surface fires. However, convection is the main heat transfer mechanism. Crown fires a more difficult to control than surface. The first accepted method for prediction of crown fires was given by Rothermal [1] and Van Wagner [2]. The semi-empirical models [1, 2] allow to obtain a quite good data of the forest fire rate of spread as a function of fuel bulk and moisture, wind velocity and the terrain slope. But these models use data for particular cases and do not give results for general fire conditions. Also crown fires initiation and hazard have been studied and modeled in detail [3–9].The discussion of the problem of modeling forest fires is provided by a group of co-workers at Tomsk University (Grishin [10], Grishin and Perminov [11–13], A mathematical model of forest fires was obtained by Grishin [10] based on an analysis of known and original experimental data [9, 14], and using concepts and methods from reactive media mechanics. Using this approach, a number of problems of mathematical modeling of forest fires were solved for the first time [9–13]. The physical two-phase models used in [15] may be considered as a development and extension of the formulation proposed by Grishin [10]. However, the investigation of crown fires initiation has been limited mainly to cases studied of forest fires propagation without take into account the mutual interaction of crown forest fires with different obstacles (roads, glades and etc.). This paper's purpose is to demonstrate the influence of these firebreaks on crown forest fires spread.

2. Physical and mathematical model of crown forest fire

It is assumed that the forest during a forest fire can be modelled as 1) a multi-phase, multi-storeyed, spatially heterogeneous medium; 2) in the fire zone the forest is a porous-dispersed, two-temperature, single-velocity, reactive medium; 3) the forest canopy is supposed to be non - deformed medium (trunks, large branches, small twigs and needles), which affects only the magnitude of the force of resistance in the equation of conservation of momentum in the gas phase, i.e., the medium is assumed to be quasi-solid (almost non-deformable during wind gusts); 4) let there be a so-called "ventilated" forest massif, in which the volume of fractions of condensed forest fuel phases, consisting of dry organic matter, water in liquid state, solid pyrolysis products, and ash, can be neglected compared to the volume fraction of gas phase (components of air and gaseous pyrolysis products); 5) the flow has a developed turbulent nature and molecular transfer is neglected; 6) gaseous phase density doesn't depend on the pressure because of the low velocities of the flow in comparison with the velocity of the sound.

Let the point x_1, x_2 , x_3= 0 is situated at the centre of the surface forest fire source at the height of the roughness level, axis $0x_1$ directed parallel to the Earth's surface to the right in the direction of the unperturbed wind speed, axis $0x_2$ directed perpendicular to $0x_1$ and axis $0x_3$ directed upward (Figure 1).

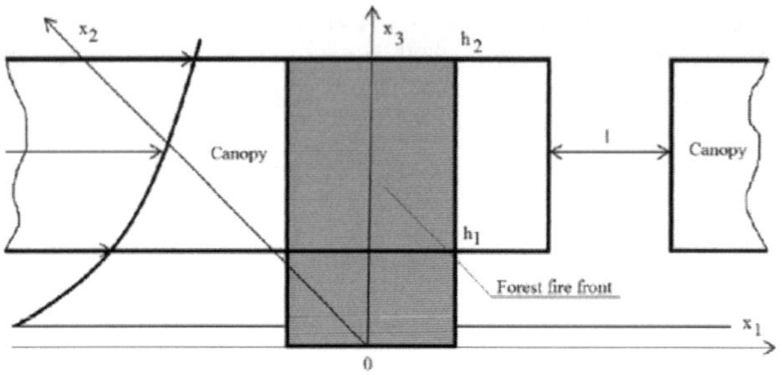

Figure 1. Computational domain.

Problem formulated above reduces to the solution of systems of equations (1)-(8):

$$\frac{\partial \rho}{\partial t}+\frac{\partial}{\partial x_j}(\rho v_j)=\dot{m},\ i=1,2,3;\ j=1,2,3; \tag{1}$$

$$\rho\frac{dv_i}{dt}=-\frac{\partial P}{\partial x_i}+\frac{\partial}{\partial x_j}(-\rho\overline{v_i'v_j'})-\rho s c_d v_i|\vec{v}|-\rho g_i-\dot{m}v_i; \tag{2}$$

$$\rho c_p\frac{dT}{dt}=\frac{\partial}{\partial x_j}(-\rho c_p\overline{v_j'T'})+q_5R_5-\alpha_v(T-T_S)+k_g(cU_R-4\sigma T^4)[0,1] \tag{3}$$

$$\rho\frac{dc_\alpha}{dt}=\frac{\partial}{\partial x_j}(-\rho\overline{v_j'c_\alpha'})+R_{5\alpha}-\dot{m}c_\alpha\ ,\ \alpha=\overline{1,3}; \tag{4}$$

$$\frac{\partial}{\partial x_j}\left(\frac{c}{3k}\frac{\partial U_R}{\partial x_j}\right)-kcU_R+4k_S\sigma T_S^4+4k_g\sigma T^4=0, k=k_g+k_S; \tag{5}$$

$$\sum_{i=1}^{4} \rho_i c_{pi} \varphi_i \frac{\partial T_S}{\partial t} = q_{3w} R_{3w} - q_{2R_{2s}} - k(cU_R - 4\sigma T_S^4) + \alpha_V (T - T_S); \quad (6)$$

$$\rho_1 \frac{\partial \varphi_1}{\partial t} = -R_{1s}, \rho_2 \frac{\partial \varphi_2}{\partial t} = -R_{2s}, \rho_3 \frac{\partial \varphi_3}{\partial t} = \alpha_C R_{1s} - \frac{M_C}{M_1} R_{3w}, \rho_4 \frac{\partial \varphi_4}{\partial t} = 0; \quad (7)$$

$$\sum_{\alpha=1}^{5} c_\alpha = 1, P_e = \rho_5 RT \sum_{\alpha=1}^{5} \frac{c_\alpha}{M_\alpha}, \vec{v} = (v_1, v_2), \vec{g} = (0, g) ;$$

$$\dot{m} = (1 - \alpha_c) R_1 + R_2 + \frac{M_c}{M_1} R_3; \quad (8)$$

$$R_{51} = -R_5 - \frac{M_1}{2M_2}{}' R_{52} = \nu_g (1 - \alpha_c) R_1 - R_5, R_{53} = 0,$$

where the symbol $\frac{d}{dt}$ marked complete derivative: $\frac{d}{dt} = \frac{\partial}{\partial t} + v_1 \frac{\partial}{\partial x_1} + v_2 \frac{\partial}{\partial x_2} + v_3 \frac{\partial}{\partial x_3}$. To determine the rates of reaction of pyrolysis, evaporation, combustion of coke, and volatile products of pyrolysis used formula:

$$R_1 = k_1 \rho_1 \varphi_1 \exp\left(-\frac{E_1}{RT_s}\right), R_2 = k_2 \rho_2 \varphi_2 T_s^{-0.5} \exp\left(-\frac{E_2}{RT_s}\right), R_3 = k_3 \rho \varphi_3 s_\sigma c_1 \exp\left(-\frac{E_3}{RT_s}\right);$$

$$R_5 = k_5 M_2 \left(\frac{c_1 M}{M_1}\right)^{0.25} \frac{c_2 M}{M_2} T^{-2.25} \exp\left(-\frac{E_5}{RT}\right),$$

where $R_1 \div R_5$, $R_{5\alpha}$ are mass rates pyrolysis forest combustible materials, evaporation, combustion of condensed and volatile products of pyrolysis and α – components gas and dispersion phase; c_{pi}, ρ_i, φ_i are specific heat, density and volume fraction of i – phase (1 is dry organic substance, 2 is water in liquid-water-drip condition, 3 are condensed products of pyrolysis, 4 is a mineral part, 5 is a gas phase); T, T_S are temperature of gas and condensed phases, c_α are mass concentrations (α = 1 for oxygen, 2 is CO, 3 are inert components of air); p is pressure; U_R is radiation energy density; σ is Stefan-Boltzmann constant; k is coefficient attenuation of radiation; k_g, k_S are absorption coefficients for gas and condensed phases; α_V is exchange ratio, phase q_i, E_i, k_i are thermal effects, activation energy and constants for pyrolysis, evaporation, combustion of coke and volatile products of pyrolysis reactions; s_σ are specific surface element forest combustible materials; M_α, M_c, M are molecular weight of individual components in the gas phase, carbon and air mixture; S, C_d are specific surface phytomass forest canopy and an empirical factor; c is light speed; v_i is projection velocity on an axis x_i; α_C, ν are coke number and fraction of combustible gases in the mass of volatile products of pyrolysis; $\dot{m}$ is mass rate of gas and dispersion phase formation; g is acceleration of gravity.

System of equations (1)–(8) describes the flow in the area of forest, which includes the space between underlying surface and the bottom edge of the forest canopy, canopy, and the space above the forest canopy. Thermodynamic, thermophysical and structural characteristics of forest combustible materials matches (FCM) of pine forest and can be found in [10–13]. Equations (2)–(4) contain the members of turbulent convection and needed closure. Turbulent stress tensor components and turbulent heat fluxes and mass are recorded via the gradients middle according to [5]. Coefficient of turbulent dynamic viscosity is defined using locally-equilibrium model of turbulence [10].

It is believed that at the beginning of time parameters of environment be considered to coincide with the values. To set up a low ground cover of grass-roots forest fire and massive speed blowing out of it. On the left edge of the wind speed is parallel to the surface of the earth and fire hearth bottom positioned at the start of the coordinate system. Then the process will be symmetrical about the coordinate plane Ox_1x_3, i.e. $x_{20}= 0$. The primary assumptions and boundary conditions for the system of equations (1)-(8) are specified as follows:

$$t=0:\ v_i=0,\ T=T_e,\ c_\alpha=c_{\alpha e},\ T_s=T_e,\ \varphi_k=\varphi_{ke},\ i=1,2,3;\ k=1,2,3;\ \alpha=\overline{1,3}. \tag{9}$$

$$x_1=-x_{1e}:\ v_1=V_e,\ v_2=0,\ \frac{\partial v_3}{\partial x_1}=0,\ T=T_e,\ c_\alpha=c_{\alpha e},\ -\frac{c}{3k}\frac{\partial U_R}{\partial x_1}+cU_R/2=0; \tag{10}$$

$$x_1=x_{1e}:\ \frac{\partial v_1}{\partial x_1}=0,\ \frac{\partial v_2}{\partial x_1}=0,\ \frac{\partial v_3}{\partial x_1}=0,\ \frac{\partial c_\alpha}{\partial x_1}=0,\ \frac{\partial T}{\partial x_1}=0,\ \frac{c}{3k}\frac{\partial U_R}{\partial x_1}+\frac{c}{2}U_R=0; \tag{11}$$

$$x_2=x_{20}:\ \frac{\partial v_1}{\partial x_2}=0,\ v_2=0,\ \frac{\partial v_3}{\partial x_2}=0,\ \frac{\partial c_\alpha}{\partial x_2}=0,\ \frac{\partial T}{\partial x_2}=0,\ \frac{\partial U_R}{\partial x_2}=0; \tag{12}$$

$$x_2=x_{2e}:\ \frac{\partial v_1}{\partial x_2}=0,\ \frac{\partial v_2}{\partial x_2}=0,\ \frac{\partial v_3}{\partial x_2}=0,\ \frac{\partial c_\alpha}{\partial x_2}=0,\ \frac{\partial T}{\partial x_2}=0,\ \frac{c}{3k}\frac{\partial U_R}{\partial x_2}+\frac{c}{2}U_R=0;\frac{1}{2} \tag{13}$$

$$x_3=0:\ v_1=0,\ v_2=0,\ \frac{\partial c_\alpha}{\partial x_3}=0,\ -\frac{c}{3k}\frac{\partial U_R}{\partial x_3}+\frac{c}{2}U_R=0,\ v_3=v_{30},\ T=T_g,\ |x_1|\le\Delta,\ |x_2|\le\Delta,$$
$$v_3=0,\ T=T_e,\ |x_1|>\Delta,\ |x_2|>\Delta; \tag{14}$$

$$x_3=x_{3e}:\ \frac{\partial v_1}{\partial x_3}=0,\ \frac{\partial v_2}{\partial x_3}=0,\ \frac{\partial v_3}{\partial x_3}=0,\ \frac{\partial c_\alpha}{\partial x_3}=0,\ \frac{\partial T}{\partial x_3}=0,\ \frac{c}{3k}\frac{\partial U_R}{\partial x_3}+\frac{c}{2}U_R=0. \tag{15}$$

Because of the horizontal sizes of forest massif more than height of forest – h, system of equations of general mathematical model of forest fire (1)-(7) was integrated between the limits from height of the roughness level from 0 to h. Besides, suppose that

$$\int_0^h \phi\, dx_3=\bar{\phi}\, h,$$

where $\bar{\phi}$ is average value of φ. The problem formulated above is reduced to a solution of the two-dimensional system of equations. It is assumed that heat and mass exchange of fire front and boundary layer of atmosphere are governed by Newton law [10].

3. Numerical solution and results

The boundary-value problem is solved numerically. In order to efficiently solve this problem in a reactive flow the method of splitting according to physical processes [11–13] was used. A discrete analogue was obtained by means of the control volume method using the SIMPLE like algorithm [16].

Fields of temperature, velocity, component mass fractions, and volume fractions of phases were obtained numerically. As a result of heating of forest fuel elements of crown, moisture evaporates, and pyrolysis occurs accompanied by the release of gaseous products, which then ignite and burn away in the forest canopy. At the moment of ignition, the gas combustible products of pyrolysis burn away, and the concentration of oxygen is rapidly reduced. The temperatures of both phases reach a maximum value at the point of ignition. The ignition processes is of a gas - phase nature.

Figures 2 present the distribution for different instants of time for temperature *a)* $\bar{T}\ (\bar{T}=T/T_e, T_e=300K)$ (1 – 1.5; 2- 2; 3 – 2.6; 4 – 3; 5 – 3.5; 6 – 4) for gas phase, *b)* oxygen $\bar{c}_1$

(1 – 0.1; 2 – 0.5; 3 – 0.6; 4 – 0.7; 5 – 0.8; 6 – 0.9), *c)* volatile combustible products of pyrolysis $\bar{c}_2$ concentrations (1 – 1; 2- 0.1; 3 – 0.05; 4 – 0.01) ($\bar{c}_\alpha = c_\alpha / c_{1e}, c_{1e} = 0.23$) for different instants of time.

The isotherms are moved in the forest canopy and deformed by the action of wind (Figure 2*a*). Similarly, the fields of component concentrations oxygen (Figure 2*b*) are deformed. It is concluded that the forest fire begins to spread. The results of calculation give an opportunity to evaluate critical condition of the forest fire spread, which allows applying the given model for preventing fires. It overestimates the rate of crown forest fire spread that depends on crown properties: bulk density, moisture content of forest fuel and etc.

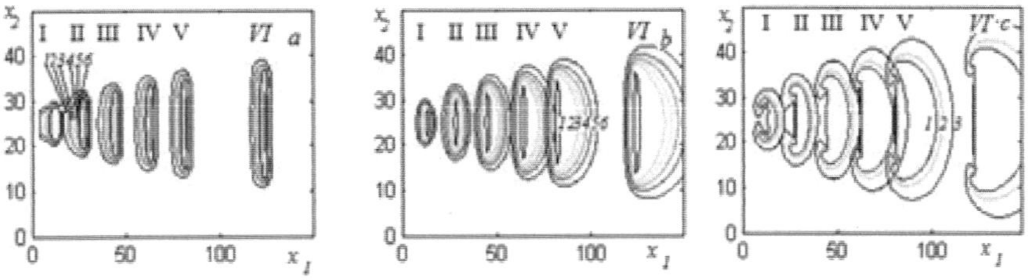

Figure 2. The distribution of temperature (*a*), concentration of oxygen (*b*) and gas product of pyrolysis (c), where I is at t=3 s, II is at 7 s, III is at 12 s, IV is at 18 s, V is at 24 s, VI is at 38 s; *V*=5m/s.

The model proposed there give a detailed picture of the change in the velocity, temperature and component concentration fields with time, and determine as well as the influence of different conditions on the crown forest fire initiation. It is important to study the interaction of forest fire front with firebreak of finite size (glade) (Figure 3*a)* temperature *T* for gas phase, *b)* oxygen $\bar{c}_1$, *c)* volatile combustible products of pyrolysis $\bar{c}_2$ concentrations).

The distance between forest fire source and glade equals 84 m. Figures 3*a, b, c* show the results of numerical simulation of a forest fire spreading around the glade under the action of wind blowing through it at a speed 5 m/s in the direction of the Ox_1-axis. Initially, the source of the fire has the shape of a rectangular. Then isotherms are deformed under the action of wind and the contour of forest fire is look as crescent. When the fire (isotherms II in Figure 3*a*) moves around the forest glade it is divided in two parts. But after that two fire fronts were joined in united fire (isotherms VI in Figure 3*a*).

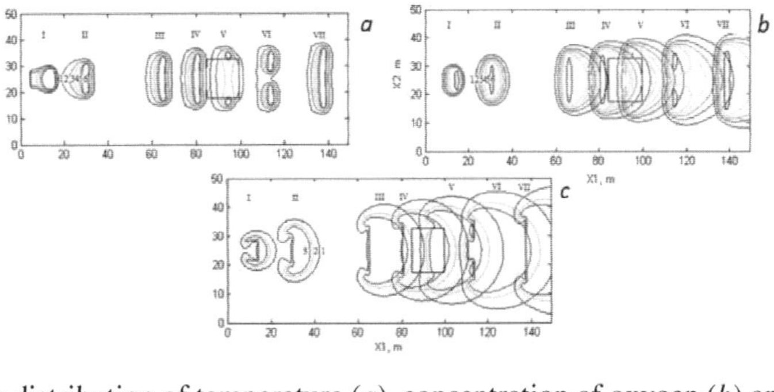

Figure 3. The distribution of temperature (*a*), concentration of oxygen (*b*) and gas product of pyrolysis (c), where I is at t=3 s, II is at 7 s, III is at 12 s, IV is at 18 s, V is at 24 s, VI is at 32 s, VII is at 38 s; *V*=5m/s.

Figures 3*b, c* present the distribution of concentration of oxygen and volatile combustible products of pyrolysis $\bar{c}_2$ for this case. If in our case the distance between the initial forest fire source and glade is increased to 74 m the crown forest fire does not spread around the glade and the forest fire dies before this clearing (Figure 4*a)* temperature $\bar{T}$ for gas phase, concentrations for *b)* oxygen $\bar{c}_1$, *c)* volatile combustible products of pyrolysis $\overline{c}_2$ concentrations). The results of calculation give an opportunity to evaluate critical condition of the forest fire spread, which allows applying the given model for preventing fires.

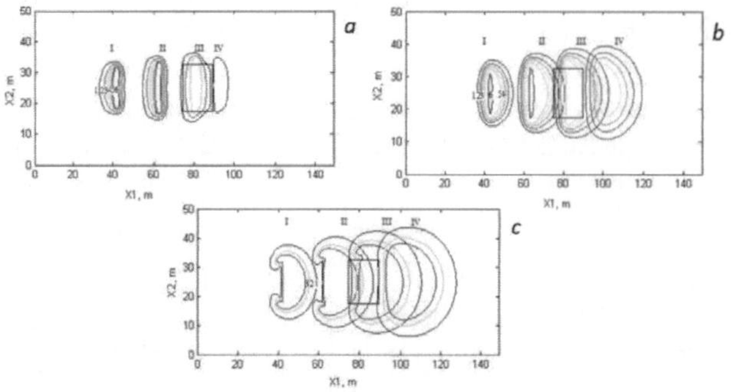

Figure 4. The distribution of temperature (*a*), concentration of oxygen (*b*) and gas product of pyrolysis (c), where I is at t=11 s, II is at 18 s, III is at 24 s, IV is at 28 s; *V*=5m/s.

4. Conclusion

The model proposed here gives a detailed picture of the change in the temperature and component concentration fields with time, and determine as well as the influence of different conditions on the crown forest fire spreading for the case of inhomogeneous of forest combustible materials supply distribution in the area and there being such obstacles to fire spread as roads, fire breaks, glades, water bodies etc. The results of calculation of the rate of crown forest fire spread obtained agree with the laws of physics and experimental data for the cases when the forest fire spread through the crowns of forest under the influence of wind velocity.

References

[1] Rothermel R 1991 *USDA Forest Service, Intermountain Research Station*, INT-438
[2] Van Wagner C E 1977 *Can J For Res* **7(1)** 23–34 doi: 10.1139/x77-004
[3] Alexander M E 1979 *Proc of the 10th Conference on Fire and Forest Meteorology: Ottawa, Ontario* 207–212
[4] Xanthopoulos G 1990 Development of a wildland crown fire initiation model *PhD Thesis* University of Montana: USA, 152 p.
[5] Van Wagner C E 1993 *Can J For Res* **23(3)** 442–449 doi: 10.1139/x93-062
[6] Cruz M G et al 2002 *Proc of the IV Conference on Fire and Forest* Meteorology: *Luso-Coimbra, Portugal* 1–11 ISBN: 90-77017-72-0
[7] Albini F A *et al* 1995 *Int J Wildland Fire* **5** 81–91 doi: 10.1071/WF9950081
[8] Scott J H *and* Reinhardt E D 2001 *USDA Forest Service, Rocky Mountain Research Station,* RMRS-RP-29
[9] Odnokopylov I G *et al* 2017 *IOP Conf. Ser.: Mater. Sci. Eng.* **189(1)** 012010 doi: 10.1088/1757-899X/189/1/012010
[10] Grishin A M 1997 *Mathematical Modelling Forest Fire and New Methods Fighting Them* (Tomsk, TSU Publishing House)
[11] Grishin A M *and* Perminov V A 1990 *Combustion, Explosion, and Shock Waves* 644–651 doi: 10.1007/BF00786501
[12] Grishin A M *and* Perminov V A 1993 *Fiz Goreniya Vzryva* **29(6)** 8–14
[13] Grishin A M *and* Perminov V A 1996 *Fiz Gorenia Vzryva* **32(5)** 107–115
[14] Konev E V 1977 *Mathematical Modelling Forest Fire and New Methods Fighting Them The physical foundation of vegetative materials combustion* (Novosibirsk, Nauka)
[15] Morvan D *and* Dupuy J L 2004 *Combust. Flame* **138** 199–210
[16] Patankar S V 1981 *Numerical Heat Transfer and Fluid Flow* (Hemisphere Publishing Corporation, New York)

Ultrawideband combined antenna with improved matching

E V Balzovsky, Yu I Buyanov, V I Koshelev and E S Nekrasov
Institute of High Current Electronics SB RAS, Tomsk, Russia
E-mail: koshelev@lhfe.hcei.tsc.ru

Abstract. To study the immunity of electronic equipment to ultrawideband irradiation, a combined antenna with improved matching has been created. In contrast to the previously presented, a novel antenna has a modified input node with a flat part instead of a cylindrical one. As a result of optimizing the geometry of the antenna electrodes, a matching band with the feeder of 0.38-2.4 GHz was achieved by the VSWR = 2. The results of the investigations of antenna characteristics in the frequency domain, as well as the waveforms of the radiated short ultrawideband pulses are presented.

1. Introduction

To test electronic devices for immunity to high-power pulsed electromagnetic irradiation, compact ultrawideband (UWB) radiation sources are required. The sources should create high intensity of the pulsed electromagnetic field at a short distance from the antenna system. The task of protecting of control circuits and communication equipment from such impacts is equally important.

Commercial devices designed to immobilize modern cars due to the failures in the operation of electronic engine management systems are well known [1, 2]. They can be used to cause a failures in the control, communication and navigation systems of drones and unmanned vehicles. The sources of high-power pulses based on the IRA antenna presenting a high-voltage switch placed in the focus of a parabolic reflector are described in [3]. Relative simplicity is distinguished by the sources having a dipole feed antenna in the focus of a circular parabolic reflector [4]. A feed antenna in the form of a TEM-horn located near the focal point of an offset reflector is noted for relative simplicity too.

To develop the antenna system of a high-power source of nanosecond pulses based on a combined antenna [5] is a subject of interest. A distinctive feature of this type of antennas is their small electrical dimensions allowing to use them as elements of the antenna array [6]. Since the combined antennas have a pattern close to the cardioid, this gives a possibility to use them as an irradiator of the parabolic reflector. For excitation of a high-power UWB source, it is preferable to use a bipolar voltage pulse since it has a mean value close to zero. This means that the spectrum of bipolar pulses has low energy of the low-frequency region, which substantially increases the energy efficiency of the radiator.

Common used communication and control protocols typically use frequencies from 0.4 to 2.5 GHz, which corresponds to the spectrum of 1-ns bipolar voltage pulse exciting the antenna. The maximum energy is concentrated near central frequency that is close to 1 GHz. Previously developed combined antennas for the frequency range indicated above, have been optimized to increase the peak field strength in the direction of the main maximum of the pattern [7]. However, the frequency band in which the antenna was matched to the feeder by the voltage standing wave ratio (VSWR) level of 2 was 0.38-1.8 GHz. The aim of this study is an attempt of further improvement of geometry of the electrodes of combined antenna for improve the antenna matching band without degrading the directional characteristics.

2. Antenna design

The antenna (Figure 1) is a combination of an electric type radiator 1, an active magnetic type radiator 2, and passive elements 3 and 4 of magnetic type. Line 5 feeding an electric type radiator has a variable cross section for creating the necessary amplitude-phase current distribution at the antenna electrodes. The antenna has a 50-ohm coaxial input 6 of increased cross-section for preventing an electrical breakdown by the voltage pulse with increased amplitude. To prevent electrical breakdown of air near the edges of antenna electrodes, the gaps between electrodes are increased too. The input element 8 has a flat part at the top to improve the matching of the antenna with the feeder in a wide frequency band. To measure the antenna characteristics, a conical adapter with an N-type connector is used.

3. Antenna characteristics

The results of measuring the VSWR of the antenna are shown in Figure 2. The VSWR was measured using an Agilent N5227A network analyzer. By the level of VSWR = 2, the antenna matching band is equal to 0.38-2.4 GHz. The antenna has the best matching with the feeder at the frequencies about 1 GHz. The maximum energy of the spectrum of bipolar pulses of 1 ns duration is concentrated near this frequency. A full-wave electromagnetic modelling of the combined antenna was made by developed software [8] based on the finite-difference in time domain (FDTD) technique. There is a reasonably good agreement between the simulated (curve 1 in Figure 2) and measured (curve 2) VSWR versus frequency. The difference between the curves 1 and 2 may be caused by the fact that the losses in the dielectric and metal neglected in the model.

In the direction of the main lobe of the pattern, the antenna radiation has a linear polarization. The plane of the polarization coincides with the plane yz. The simulated antenna patterns in the E- and H-planes are shown in Figure 3 and Figure 4 (curves 1). The main lobe of antenna pattern remains the direction along the z axis and the main lobe has no damping exceeding 3 dB in the frequency band equal to 0.3–3 GHz.

Figure 1. Antenna design.

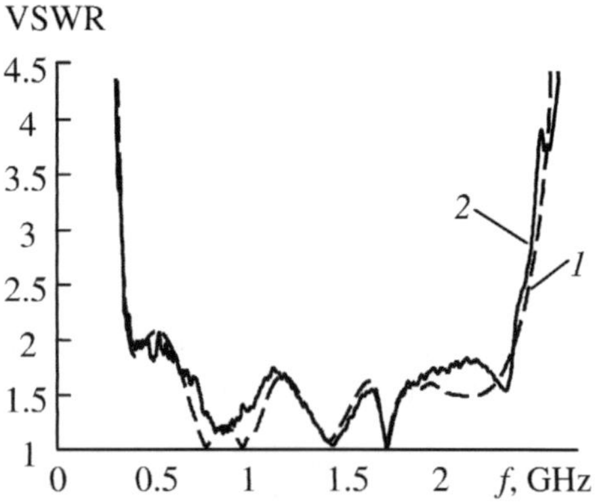

Figure 2. VSWR of the antenna: 1 – simulations, 2 – measurements.

Measurement of radiation patterns was carried out in the time domain. A low-voltage 1-ns bipolar pulse generator was used. The pulses were recorded with a LeCroy Wavemaster 830Zi digital storage oscilloscope. At the output of an auxiliary UWB receiving antenna, the pulse waveform was recorded for each angular position. Using the Fourier transform, we obtained the angular dependences of the spectral components in the 0.3-3 GHz range. Thereby plotted antenna patterns in the E- and H-plane are shown in Figure 3 and Figure 4 (curves 2).

A TEM-antenna with the ground plate dimensions of 120×50 cm and an aperture height of 8 cm was used as a standard receiving antenna for measuring UWB pulse waveforms. The voltage at the

TEM-antenna output is proportional to the incident field strength, and the effective length in a wide range is independent of frequency and equals to half the aperture height. The waveform of the voltage pulses at the output of the TEM-antenna located at a 5-m distance from the transmitting one is shown in Figure 5. The simulated waveform of the electric field strength at the same distance is presented in

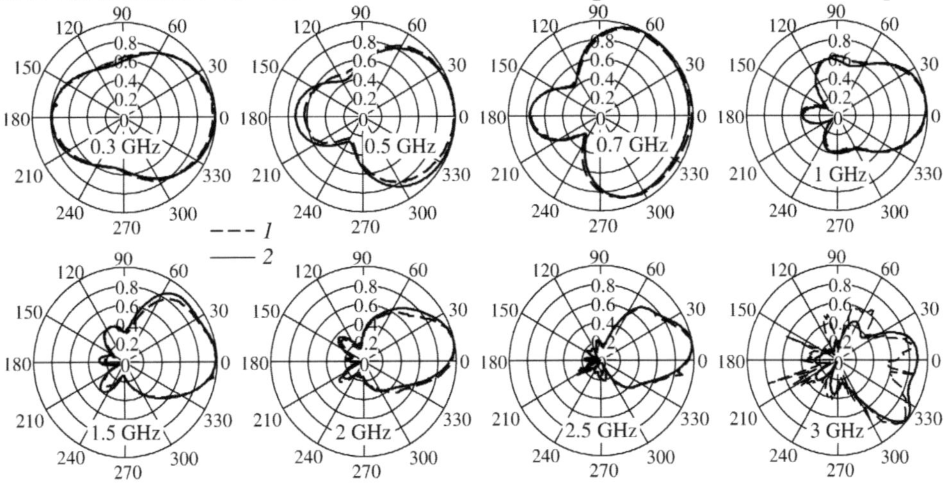

Figure 3. Antenna patterns in the *E*-plane at different frequencies: 1 – simulations, 2 – measurements.

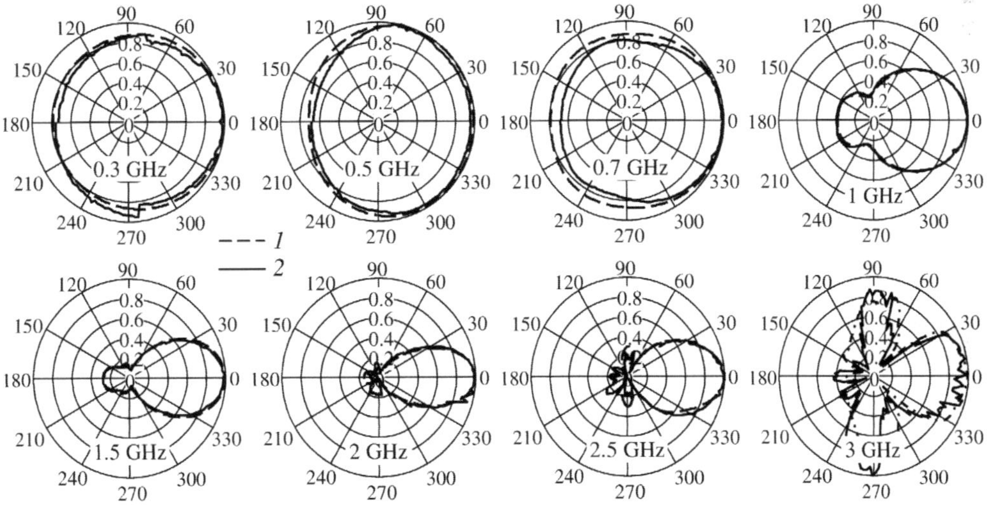

Figure 4. Antenna patterns in the *H*-plane at different frequencies: 1 – simulations, 2 – measurements.

Figure 5 as well. The difference between the curves of simulated and measured patterns in the range of 4-5 ns may be caused by the reflections from the support carrier of the receiving antenna, or from the outside conductor of the coaxial cable behind the receiving antenna.

The waveform of the radiated pulses depends on direction of observation. To quantify the value of distortion of the waveform of the pulses, the root-mean-square (RMS) deviation σ of the pulses *V(t)* and *U(t)* was calculated according to the expression: $\sigma = \sqrt{\int_T \left[v(t) - u(t-t_0)\right]^2 dt \Big/ \int_T u^2(t)dt}$. Here, *V(t)* is a voltage pulse at the output of the receiving antenna, when the transmitting antenna was rotated to an arbitrary angle θ relative to the axis z. *U(t)* is the voltage at the output of the receiving antenna in the direction $\theta = 0$. Values $v(t) = V(t)\Big/\sqrt{\int_T V^2(t')dt'}$ and $u(t) = U(t)\Big/\sqrt{\int_T U^2(t')dt'}$ are the normalized functions, T is the time window, and t_0 is the time shift of $U(t)$ relative to $V(t)$, at which σ drops to the minimum value. The measured RMS deviations of the pulse waveforms at the output of the TEM-antenna at $0 < T < 4$ ns are shown in Figure 6 in the E-plane (curve 1) and in the H-plane (curve 2). When $\sigma < 0.05$, the two functions $V(t)$ and $U(t)$ are practically indistinguishable. The difference of two waveform samples of the same pulse generator registered by the real-time oscilloscope without averaging at different points of time is estimated by the value of $0.05 < \sigma < 0.1$. Appearance of an additional time lobe on the waveform leads to increase of σ to the value of approximately equal to 0.3. For $\sigma > 0.6$, the two waveform samples differ significantly from each other. The angular sector in which σ is no higher than 0.2 has the value of $-35 < \theta < 60$ degrees in the E-plane and $-40 < \theta < 40$ degrees in the H-plane.

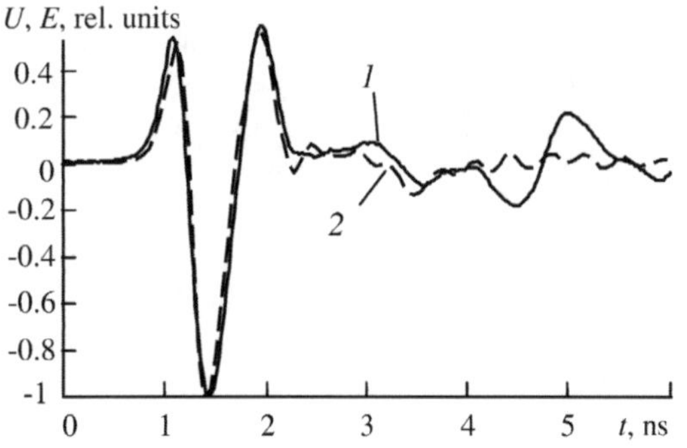

Figure 5. Measured waveform of voltage pulse at the output of receiving TEM-antenna (1); simulated waveform of the electric field strength at a 5-m distance from the antenna (2).

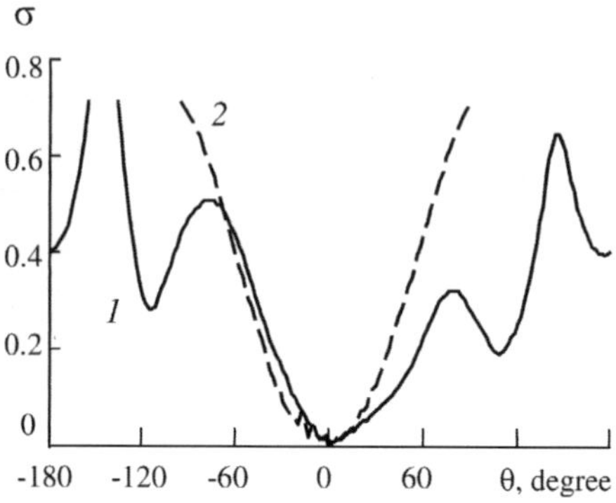

Figure 6. RMS deviation of the pulse waveforms at $\theta = 0$ and arbitrary θ at E- (1) and H-plane (2).

4. Conclusion

A combined antenna with improved matching with a 50-ohm coaxial feeder has been developed. The antenna has a modified input node with a flat part instead of a cylindrical one. Resulting from optimization of geometry of the antenna electrodes, we achieved a 0.38-2.4-GHz band of matching with feeder by the VSWR level of 2. The bandwidth at which the maximum of the antenna pattern retains its position without damping by more than 3 dB is not less than 0.3-3 GHz. The antenna is designed to radiate UWB pulses with a spectrum in the frequency band 0.4-2.4 GHz, or narrowband signals within this frequency band. The increased gaps between the antenna electrodes allow radiating UWB pulses with increased voltage for testing electronic equipment for electromagnetic compatibility.

References

[1] Urbancokova H, Valouch J, Kovar S 2015 *Przeglad Elektrotechniczny* **91** 101
[2] Hong K, Braidwood S 2010 *Proc. Int. Conf. Electromagn. Adv. Appl.* 378
[3] Baum C E, Farr E G 1993 *Ultra-wideband, Short Pulse Electromagnetics* (New York, Plenum Press)
[4] Ryu J, Lee J, Chin H, Yeom J H, Kim H T, Kwon H O, Han S H, Choi J S 2013 IEEE *Trans. on Plasma Sc.* **41** 2283
[5] Andreev Yu A, Buyanov Yu I, Koshelev V I 2005 *J. Commun. Technol. Electron.* **50** 535
[6] Efremov A M, Koshelev V I, Kovalchuk B M, Plisko V V, Sukhushin K N 2014 *Laser and Particle Beams* **32(3)** 413–418 doi: 10.1017/S0263034614000299
[7] Koshelev V I, Plisko V V 2017 *J. Commun. Technol. Electron.* **62 (6)** 565–568 doi: 10.1134/S1064226917050096
[8] Zorkaltseva M Yu, Koshelev V I, Petkun A A 2017 *Rus Phys. J.* **60 (8)** 1291–1297 doi: 10.1007/s11182-017-1210-8

Development of expert systems for analyzing electronic documents

AL-AZZAWI Abeer Yassin[1,2], S Shidlovskiy[1], A A Jamal[2]
[1]National Research Tomsk State University, Tomsk, Russia
[2]General Directorate for education of Diyala, Diyala, Iraq
Email: abeerda89@gmail.com

Abstract. The paper analyses a Database Management System (DBMS). Expert systems, Databases, and database technology have become an essential component of everyday life in the modern society. As databases are widely used in every organization with a computer system, data resource control and data management are very important [1]. DBMS is the most significant tool developed to serve multiple users in a database environment consisting of programs that enable users to create and maintain a database. This paper focuses on development of a database management system for General Directorate for education of Diyala in Iraq (GDED) using Clips, java Net-beans and Alfresco and system components, which were previously developed in Tomsk State University at the Faculty of Innovative Technology.

1 Introduction

Since 1960, especially in the developed countries, most businesses began to store and save files on the computer; the experts of computing began to think about the development of theories and methods for re-using files that are stored in the computer. These files are called computerized.

The Department of employees' affairs (DEA) is one of the important sections in GDED and is a measure of the level of management in the organization. Some routine and burdensome types of work need accuracy and cannot contain errors, for example, in basic information for employees, work management etc. Staff specialists in this department spend much time and make great efforts while managing these actions manually [2]. There are unmatched features through the use of computers to manage the information obtained from applications and documents of people who want to get a job in GDED. These features include the speed of search, security, large memory space, long life, low cost, great reliability and proper search, which can improve the effectiveness of employee information management [3].

2 Case study: Department of employees Affairs (DEA)

The DEA is being the most important administrative department in any organization because of being involved in relations with all organization's employees. They are the basis of the work process, so the advancement of employees' affairs leads to the advancement of the whole organization.

The methods adopted for the management of this department are manuals represented by records and files relating to the information of employees, as shown in Figure 1. Every employee in GDED has a dossier containing all the information and its own administrative orders, CV from the moment of admission, details about attendance and absence at the working place and other types of information [4].

Figure 1. Dossiers of the employees used in the department.

2.1 Problems with the old methods used in DEA

Several problems can be summarized because of using old and traditional methods:

- Delays in the delivery of services because of the manual work.
- The department requires an adequate number of staff in order to make its administration because of existing manual employee's information system.
- Information and data stored in several places lead to delay and inaccuracies in the task completion.

2.2 The Purpose of the Study

The main purpose of this study is to design database for constructing an employees' information management system in the department of employees' affairs in GDED. This system will solve most of the problems and obstacles resulting from the old methods and techniques that have been previously mentioned. This will allow transforming the work in this department from manually to a computer based-system, which, in turn, will improve accuracy, efficiency, security and so on. The problems arising from the purpose include the following:

- How to reduce the errors and inaccuracies in the process of managing employee's information.
- How to reduce the time and efforts required to do routine operations of the DEA.
- How to support the administration of the organization in the decision-making process.

2.3 Implementation of the system design

After setting goals and vision, for which the SIS was designed, we implemented the process of the system design in order to meet the needs of the GDED. The system represents a system for processing employees' data and organizing it in the form of useful information for the management of the organization. The system allows improving the management process in employees' affairs department and increasing speed of operations completion with high efficiency and low errors. An information system (IS) can be defined as a set of interconnected components that collect, process, store and distribute information to support decision-making, organizing and control in an organization. Furthermore, information systems may also help managers and workers to analyze problems, imagine complex subjects and create new products. The general model of an IS is in the form of a collection of semantic services.

Information systems contain information about important people, places and things within the organization or in the surrounding environment. Information is the data that was designed into a form that is meaningful and useful to human beings. Data, in contrast, are streams of raw facts representing events occurring in organizations or the physical environment before they have been organized into a form that people can understand and use as shown in Figure 2.

Therefore, the primary function of information systems is to hold a series of operations on the data and turn them into information in order to help decision-makers in the organization to take the proper decision as soon as possible.

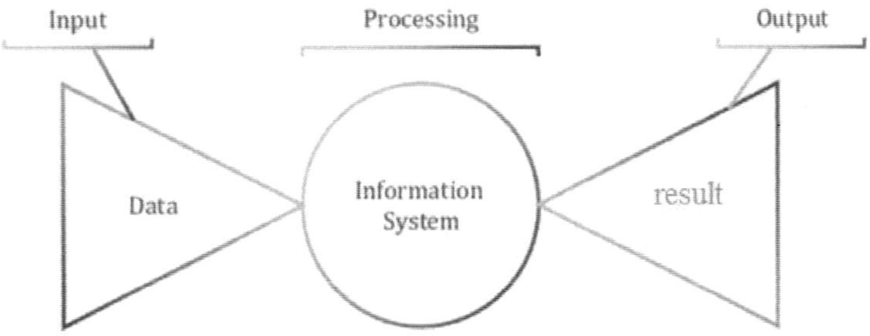

Figure 2. Information system.

3 Documents and requests receiving by using (Alfresco)

People who want to apply to get a job in GDED, have to fill down all the required information using Alfresco. Then, through NetBeans, Alfresco will send information about the person's qualifications to Clips, which, in turn will analyze the person's qualifications and make a decision (Figure 3).

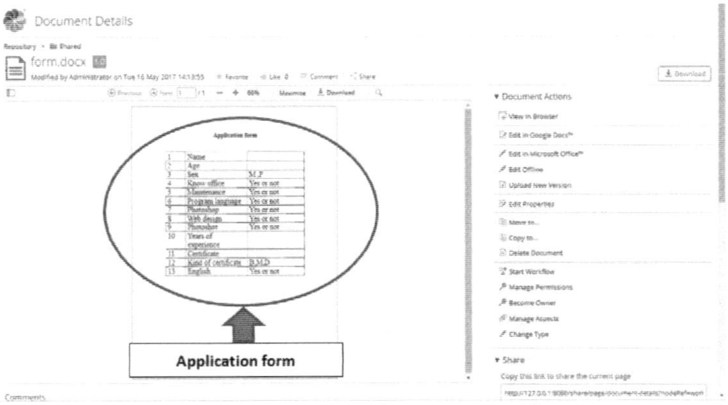

Figure 3. Application form using Alfresco.

4 Data analysis using Expert system (Clips)

In order to enable the DBMS to classify requests from people who want to get a job in the organization and send the request to specific department according to person's skills, we have to determine the requirements of each department, such as person's specifications, skills, experience and academic achievement, etc. We determine all these things using Clips and java NetBeans by writing the programming code in Clips system.

5 Making API for the system

To make an Application Programming Interface (API) we need to connect Java with Clips, therefore, we use NetBeans as shown in the Figure 4.

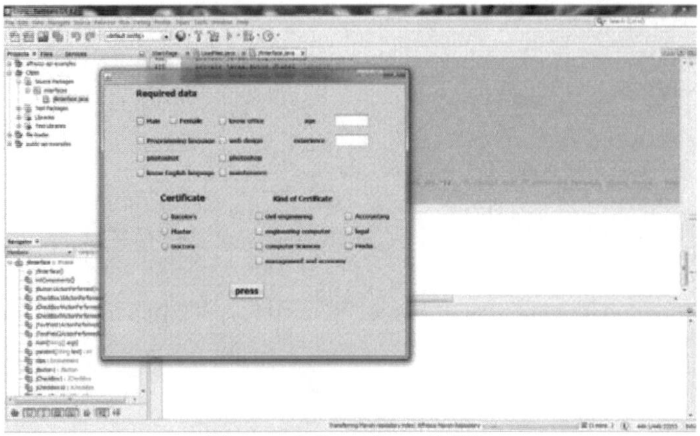

Figure 4. Interface (API).

6 System working

After filling the person's qualifications, the system will analyze these qualifications and choose the perfect department for the person to work in (Figure 5).

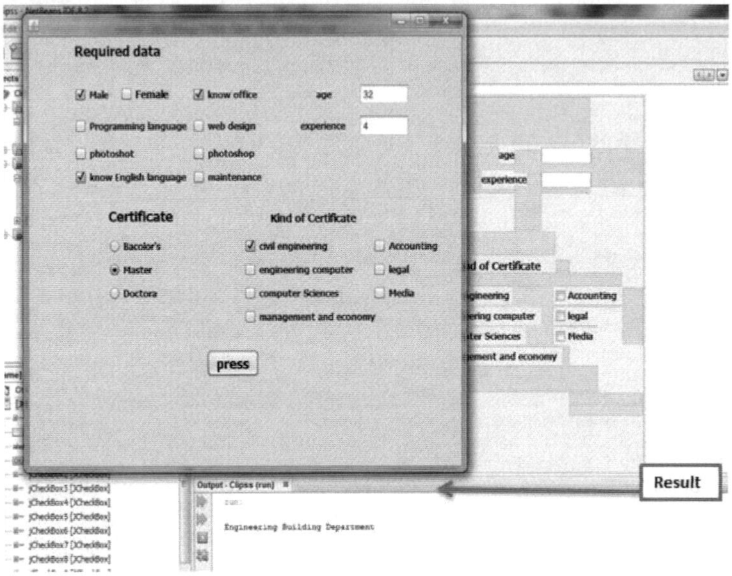

Figure 5. System interface and result.

In the case presented in Figure 5, the system has chosen the engineering building department for the person.

7 Conclusion

In this paper, we have introduced a database management system for employees in GDED. The main objective of designing and carrying out this system is to get the possibility to collect employees' data in one place, make various transactions and modifications on it, and retrieve this data according to requirements of the organization. This objective as well as the desired system functions were achieved. This system will transfer the manual work in this department to the computer-based system. In addition, this system can promote the management of organization to carry out its tasks more effectively. The Employees Information System can reduce time and efforts required in the process of managing employee' information as well as the need for a large number of staff to carry out the functions of the DEA because managing of this system needs to only one employee. Moreover, it will help the organization's administration to speed decision-making because the system allows using real-time data when making decisions, which benefits to the organization.

The system presented in this paper needs to be in service for the transition phase.

The further development of the system includes plans to link the employee Information Management System (EIMS) with the GDED website. This will enable people who want to apply for a job in this organization to apply directly through the online interface embedded on the GDED website and make requests online, thus, reducing time processing.

Acknowledgments
This study (research grant No № 8.2.24.2018) was supported by the Tomsk State University competitiveness improvement programme.

References

[1] Rumyantseva T B, Syryamkin M V, Syryamkin V I , Vaganova E V 2016 *Technology Management. Part 3. Technology Project Management: textbook* (Tomsk, STT)
[2] Introduction to Computer Information Systems / Database URL: http://www.webopedia.com – lang.Eng
[3] Shimozono K, Itsukiy M, Harasaka Y and Furukawa Z 2010 *User Management in an Educational Computer System: Personal Information Management* (Fukuoka, Japan)
[4] General Directorate for education of Diyala URL: http://www.diy.epedu.gov.iq/

IOP Conf. Series: Materials Science and Engineering 363 (2018) 012004 doi:10.1088/1757-899X/363/1/0120

Educational robotics as an Innovative teaching practice using technology: minimization of risks

S B Kvesko[1], N G Kvesko[2], A A Korniyenko[3] and N N Kabanova[3]
[1]National Research Tomsk State University, Tomsk, Russia
[2]Siberian Federal University, Krasnoyarsk, Russia
[3]National Research Tomsk Polytechnic University, Tomsk, Russia
E-mail: svetla_kvesko@mail.ru

Abstract. This research is focused on studying educational robotics, specifically robots which provide functions of educational activity. We have considered the questions of intelligent agents' behavior and have studied their educational opportunities. Educational robotics is a powerful tool of developing person's skills and abilities in various fields of technical creativity and professional activity. The evolutionary development of robotics is connected with development of artificial intelligence, where emotions play a great role in operations. Nowadays the main thing is to form the ability and skills of optimum interaction with social environment when a person, based on gained knowledge, is capable to put goals of the activity in strict accordance with laws and society conditions and using current technology.

1. Introduction

The artificial intelligence does not excel human intelligence. There are many interpretation of the concept "artificial intelligence". According to Allen Newell and Herbert Aleksander Simon, the founders of research in the field of artificial intelligence [1], it is necessary to consider the specific strategy of machine cognitive and peculiar conscious activity. These strategies cause various potential opportunities of a human, considerably strengthening not only his/her physical capacities, but intellectual as well.

David Vernon formulated the concept of computer vision and robot's capacity to self-studying, he practically raised the question about the anthropomorphous robot. In his opinion [2], progress in the sphere of developing and creating robots is too fast, that something that was considered as an artificial intelligence a few years ago is seen as an improved method of research and data processing now.

The aim of the research is to consider the essence and peculiarities of educational robotics as an innovative pedagogical practices using technology, to solve the problem of risks minimization in the usage of robotics in educational environment.

As a result, we have solved such tasks as revealing the subject of educational robotics as a pedagogical technology and forms of its implementation. To achieve the goal and do the tasks we have used cognitive and productive approach in combination with dialectic and competency-based approach. Application of educational robotics within educational process is developed in cognitive and activity plane [3]. Cognitive activity of subjects of educational process is composed of requirements and strivings to acquire knowledge with the assistance of wide range of various sciences, young researchers' (trainees) interest in a problem solving, trainees' personal educational trajectory developing, cross-fertilization between theory and practice in the course of problem solving at the intersection of areas of knowledge. We applied competence-based approach which helped us to analyze social space of safety formation as a space of organized human activity for eliminating social risks under formation of effective educational process. The competence-based approach has a complex hierarchic-level structure with a plenty of components and is interconnected with system, structural, functional, problem, genetic approaches in studying phenomena and events. This method is focused on a personality and the activity of the personality. Competence-based approach refers to the purpose and the result of education. Competence-based approach holds a special place in training specialists in information and measurement equipment and technology as it integratively opens the questions, requiring the solution, and then solves these problems integrally. Certainly, competence is a social education requirement, but this requirement is rather alienated from a future specialist. The competence is not simple knowledge acquisition; it is relevant competence, skill, knowledge of this competence, a personal quality. Cognitive and productive approach is a method of studying by autonomous system of environment and world, in general, learning by own experience, perceiving consequences of the acts, searching the ways to achieve target tasks and. objectives and adaptation to circumstances which are constantly changed.

2. Educational Robotics as an Innovative Teaching Practice

Educational robotics is a new knowledge acquisition at the intersection of the fields of priority and advanced areas of knowledge in such scientific spheres as mechanics, electronics, automatic equipment, programming, project engineering, circuitry engineering, design engineering. It represents such pedagogical practice using technology when acquisition of knowledge happens in the process of creating a robot. Therefore, educational robotics is focused on two actions: educational and technological.

These two actions are combined and are urgent in the recent period, the period of innovative economy formation and development. We need to emphasize that educational robotics primarily is based on integration of knowledge, received while studying subjects according to educational standards. However, to solve certain specific objectives or problems cognitive achievements of sciences taught variably are integrated [4]. Consequently, we deal with knowledge integration which unites both the first, and the second integrated systems. As result we get the following:

– formation of direct connection between various subjects and scientific branches of knowledge; understanding of education process, meaning of education, necessity to compete in the field of robotics;
– ability to achieve successful results during research and educational process;
– realization of the rivalry of the robotic ideas and decisions in the world.

Nowadays theoretical knowledge is popular in unity with practical knowledge. Owing to the currently central demand of both theoretical and practical the striving to master them more deeply and more widely arises. This can be reached by means of robotics as a technology for mastering a complex of knowledge. Robotics as a pedagogical practices using technology corresponds to the basic Didactic principles of education [5]:

– scientific orientation;
– world outlook orientation;
– problematicity (stating a problem, having several decisions);
– visual aids caused by robotics (drawings, schemes, designs, mechanisms);
– self-actualization of students' own needs for knowledge, self-studying the theory and the transferring theory into a practical product, that is to the robot;
– variability in studying, in choosing complexity of an engineering problem;
– sequence of learning knowledge at robot production;
– solidity and cyclicity of studying;
– stage-by-stage checking robot production;
– deepening knowledge expansion which is necessary for each stage;
– unity of the educational and developing functions of studying.

Modern scientific and technical creativity is expressed through a robotics which is focused on searching, preparing, and implementing younger generation ideas at the interface of researchers on urgent fields of knowledge with vast research prospect in the future. Therefore, robotics represents the unique innovative pedagogical practices using technology sent to search for creative persons. Robotics as pedagogical technology is expressed in educational process in the following forms [6]:

1. Seminars.
2. Symposiums, conferences.
3. Open online courses.
4. School and Conference.
5. Workshops.
6. Forums.
7. Special courses.
8. Specialized laboratories of robotics.
9. Participation in competitions.

The main objective of all forms of expression of robotics as innovative pedagogical practices using technology consists in [7]:

1. students acquaintance to the main principles of robotics, drawing attention to youth scientific and technical creativity;
2. exchange of experience between young researchers;
3. the organizations of communication between subjects of educational process;
4. discussion of the ideas and results of technical realization;
5. knowledge and basic principles of robotics;
6. development of bases of design and designing of robots of various degree of complexity;
7. understanding of an essence of technical creativity;
8. ability to work in team.

Robotics is a form of additional education at schools, lyceums, gymnasiums. In colleges, technical schools, institutes, universities, academies, it keeps the character of additional education. However in educational institutions with robotic profile, robotics is additional education along with the primary educational activity. The teachers, carrying out this form of pedagogical activity, create author's programs on organization and implementation of this innovative pedagogical practices using technology.

At the same time there is a scientific and methodical challenge connected with the problem of concretizing knowledge on robotics course having cross-disciplinary character. As a result, the special course on robotics will allow to implement the ideas of the cross-disciplinary complex, and to promote practical use of theoretical knowledge and enhanced studying of the entering this complex disciplines. Besides, it is possible to enrich robotics course using information from related areas.

Conducting lessons in the international robotics laboratories will be a considerable contribution for development of educational robotics. As a result of involving young forces in such scientific branch as robotics, there will be not only search, but also preparation and support of new highly qualified experienced staff [8]. Owing to successful search, challenging ideas, we can refer to high intelligent potential of the educational system subjects engaged in the educational robotics program. Annually competence content requirements are changed, new ideas appear constantly, and interesting decisions are made. Cognitive character of robotics is developed increasingly. Strengthening the role of educational and pedagogical technology is an important task of educational robotics - it is students' work on cross-disciplinary complex projects in the field of robotics, formation of the up to date intelligent robotics environment which is capable to train experts for developing high technologies.

Competitions, the Olympic Games, conferences having competitive character and characterized by any given degree of competitiveness are of great importance for development of educational robotics. As a rule, competitions are designed for a great pool of participants. The Olympic Games are for intellectual elite. Conferences are for broad interchange of opinions, for scientific and creative intercommunication. In competitions, the Olympic Games, conferences participants are engaged in self-studying in the field of educational robotics. They can use gained knowledge and modern technologies in practice. In these forms of educational process students gain experience in implementing an innovative project, prepare to work in the hi-tech companies in the sphere of robotics.

Robotic competitions are one of the key insights of educational robotics. They give an opportunity to disclose technical imagination, to exchange the ideas, knowledge, approaches to master knowledge. Cross-disciplinary creative approach in educational robotics is connected with implementation of symbolic method. Educational robotics demands using active and interactive training. In educational process a modeling environment should be created, in which a student works with knowledge and receives an intellectual response of system, and creates the virtual and real world. The symbolic method in educational robotics acts not in opposition to formal logic, but as a connection between real and unreal.

The issue of scientific knowledge is solved not only in harmony of forms and techniques of expressing ideas, but in visual knowledge of the scientific truth as well, in its internal reflection. Scientific ideas have to declare themselves, emotionally influencing student's consciousness. The abstract logical images gaining sensual character turn into concepts of educational robotics. The purpose consists in developing their ability to make sense of life phenomena transformations. In the system of contemporary education there is a tendency to symbolize forms of knowledge acquisition and represent it to integration of scientific, symbolic and mystic knowledge [9].

It is a basis of innovative educational technologies which allow to give knowledge in a creative and figurative form that promotes its stronger training. Symbolic method helps to implement internal human aptitude to studying. It allows to achieve a bright illustration of upbringing and education process. This method influences feelings, rather than exclusively education subject's mind via creating robotic models.

At the carried out study we have succeeded in research objects:

1. The investigation has revealed the subject and the features of educational robotics as an innovative pedagogical practices using technology.
2. We have considered the forms of its implementation.
3. We have developed methodical guidelines on implementing educational robotics as an innovative pedagogical practices using technology.

The innovative educational technology presented by educational robotics is directed to stimulate and develop scientific and technical creativity of youth. The main feature of this technology is to combine competences formulated during the continuous comprehensive educational process of receiving and forming practical skills and theoretical knowledge allowing to realize the creative potential of young people.

Educational robotics is a practical implementation of innovative approach to youth scientific and technical teaching, namely, it is a cross-disciplinary complex of the training, developing, cognitive and methodic actions aimed at the developing young inventors and researchers [10]. The main idea of educational robotics consists in:

1) integration of subjects to create robot and consistently learn all steps of engineering process - from controlled robot to autonomous robot;
2) organizing design and inventive activity;
3) providing life cycle of creative projects – from the idea to an implementation;
4) differentiation of attainment level;
5) providing team work;
6) implementation of professional orientation;
7) using contemporary high technologies.

Use of robotics is connected with risks in educational process, influencing subjects of the educational environment negatively. As a result, it was revealed that robotics and robotic technologies define depth and amplitude of the scope of knowledge. The carried out analysis of the role of social risks in educational system has shown that the process of formation and development of education depends on the state and the nature of social risks, on the extent of preventive measures with them. The study emphasizes that minimizing risks can be achieved by the tasks limitation and concentration of conditions to use educational robotics, and establish criterion and standards of its application. Orientation of education assessment results is relevant due to the necessity of the theoretical and methodological analysis of training specialist in the robotics and robotic technology. According to the applied method of the analysis and cross-disciplinary approach we conclude that educational process has an integrative nature.

3. The issue of risks and their minimization in using robotics in educational process

Innovative character of educational robotics as pedagogical technology is caused by the combination of the learners' project and practice-oriented activity and use of the modern technologies. Educational robotics is a powerful tool of developing person's skills and abilities in various fields of technical creativity and professional activity.

The key social task for modern society is to form population safety, to establish and develop tolerant relations in society and in education as well. In the modern era the state of scientific and technical assurance of educational process is closely connected with the development of innovative teaching technologies, one of which is robotics in educational process. Applying robotics in educational environment causes risks in implementing interaction of subjects and subjects and objects of educational process as well. To achieve necessary result we need to use methods of risks minimization. Applying these methods it is possible to reach minimization of risks when using robotics in educational process.

The risk inherent in application of such supporting training tool as a robot can influence the result both positively and negatively. It is possible to avoid risk. To achieve this it is enough to apply an avoidance method, that is, to refuse the action connected with risk or to develop the action which is completely excluding risk. However it is not always possible to avoid risks. Therefore, avoidance of one risk can lead to emergence of another. Hence, it is necessary to analyze various ways of minimization [11]. It is possible to reach minimization of risk by limitation of tasks and concentration of conditions for use of educational robotics, and establishment of criterion and standards of its application. The limit of set up problems for particular conditions of studying or research allows to specify the application of a certain type of robotics (specific robots) in the organization of educational process.

The subject of education loses the experience of a harmonious state if a priority and the importance of robotics. Thus, there is a self-loss of "I-ego", loss of understanding the essence of the conscious activity and its difference from robotic, underestimation of the intelligence. There is an emerging of self-estrangement, and the state of admiring robotics, self-opposition, the rise of technologies above a person. From there, the risk of losing yourself as a person, and emergence of dependence on robotics appears. According to the personality's self-loss of its "I-ego" the model of person's social regeneration can be formed. In the beginning a person, being in harmony along, considers him/herself intellectual, creative. But with the development of artificial intelligence, robotics, virtual games, people become the elements of an information field. The first prerequisites of alienated state, alienated thoughts appear. Here we discuss negative results of applying robotics in educational process.

But there are also positive results. Use of robotics promotes knowledge of the world around, transformation of the environment. In this case robotics acts as the tool, a learning tool. For users it is important to master cognitive features of educational robotics, resulting in understanding the importance of intelligence of a person, his/her creativity. Cognitive approach in formation and use of educational robotics allows to increase positivism of using innovative technologies in educational space by means of identifying the mechanism of interaction of the education and robotics subject intelligence as learning tools. The complexity is that understanding of this interaction is possible after understanding the consequences of risk. In modern society the lines of social strain exerting negative impact on education. As a result, risk situations happen [12]. Modern technology equipment, including robotics influences much on risks emergence.

In education system there was always an ideal to educate an all-round and harmoniously developed individual, being not only self-sufficient and having fundamental knowledge, but brought up in the system of moral values. Usefulness and performance indicators [13], that is education commercialization, are the main content of modern educational practice. As a result humanitarian knowledge becomes second-grade, the value of technical and other applied sciences increases. The modern youth is interested in intelligent systems now and gravitates towards developing robots and certain cloning of properties and lines of the person in the robot, believing that it is possible to replace person's monotonous work by robots.

4. Summary

Application of robotics and appropriate technologies facilitates and contributes to the development of interactivity in teaching, developing student's independence and creativity, but at the same time forms egoism and individualism. The egoism and individualism are those features of an individual, which leads to alienation, to professional burning out, emotional failures. The innovation of a robotic technology represents a form of vulnerability of a person against the threats caused by the modernization and development of robotics [14].

The risks caused by an innovation characterize the development stage when old mechanisms lose their efficiency, and new one arises [15]. Unfortunately, the influence of robotics affects often that absolutization of artificial intelligence leads to self-reflection, loss of the "I-ego".

Innovative teaching techniques using robotic technology activate students' attention and perception, make creative atmosphere. The feature of current generation is orientation to the entertaining, visualized teaching techniques specific to contemporary culture [16]. Innovative teaching is risky as it uses untested teaching methods. Teaching by robotic technology is innovative and risky as it is impossible to predict the action of the artificial intelligence. But at the same time use of a robotic technology leads to passive learning material, does not make memory and thoughts to train, as everything is done by a robot. There is time and forces saving, but the psychological indifference is formed.

Competence-based approach is of great importance in training specialists by using robotics and robotic technology. The core of training specialists in the field of robotics is competences, aimed to form system of knowledge, skills and abilities. Nowadays, breakthrough technologies take on great importance. It is necessary to emphasize the role of information and measuring competences. In the course of their mastering people generate knowledge regarding the activity in the field of intelligent systems. The modern robotechnics and robotic technology assume rather good level of computer, information, natural-science, economic and socio-humanistic training of a person. Both in life, at work, and during educational process people deal with the innovative and advanced technologies which very quickly become outdated and make the owner be in a constant stress which can lead to emotional, and even to professional burning out [17].

Communicative practice is a part of business communication. For communicative practice in the system of robotics it is necessary to have skills and knowledge in the information and measuring techniques and technologies, knowledge of using them [18]. In the implementation of communicative practice by means of robotics it is necessary to be professional in using methods of acting in various situations. Robotic technology is favorable for development of individual approaches to training future specialists. Owing to cognitive features this is a fundamental condition of a competence-based approach [19,20].

Robotic technologies provide the efficiency of technical means and their profitability via optimum distribution of information and actions [21]. The robotic technologies are adaptive. When they were introduced into educational process, then methods, technologies and a methodological support of educational process using robotic technology and robotics started regulating.

Global penetration of robot technologies into education considerably raised technical and methodical capabilities, it promoted enhancement of didactics. New quite efficient forms of education appear with the development of robotics.

Educational process using robotics and robotic technology keeps continuity of traditional education system, contributes to the development of innovative teaching model. The birth of robotics and robotic technology, despite the risks, makes changes in education system. The problem of minimizing risks in the use of a robot technology in educational process requires the solution owing to the existing education process modernization.

References

[1] Rassel S, Norvig P 2015 *Iskusstvennyy intellekt: sovremennyy podkhod* (Moscow, Williams Publ.)
[2] Jelassi T 2014 *European Casebook on Competing Through Information Technology, Strategy and Implementation* (New York, Plentice Hall)
[3] Castells M 2010 *The Rise of the Network Society* (London, Blakwell Publishers)
[4] Nonaka I 2008 *The Knowledge-creating company: How Japanese companies create the dynamics of innovation* (Oxford, Oxford University Press)
[5] Shumilov V N, Syryamkin V I, Syryamkin M V 2015 *AIP Conf. Proc.* **1688** 040007 doi: 10.1063/1.4936040
[6] Limoncelli T A, Hogan C J, Chalup S R 2007 *The Practice of System and Network Administration, 2nd Edition* (United States, Addison-Wesley)
[7] Kvesko R B, Kvesko S B, Salkova N E, Shinn T N 2005 *9th Korean-Russian International Symposium on Science & Technology–KORUS 2005* 1039 – 1092
[8] Syryamkin V I, Bogomolov E N, Kutsov M S 2014 *Advanced Materials Research* **1033-1034** 1327–1330 doi: 10.4028/www.scientific.net/AMR.1033-1034.1327
[9] Wenger E C and Snyde W M 2000 *Harvard Business Review* 78 139–145
[10] Drucker P 2011 *Innovation and Entrepreneurship (*New York: Collins)
[11] Bek U 1992 *Risk Society: Towards a New Modernity* (London: Sage)
[12] Bourdieu P, Wacquant L 1992 An invitation to reflexive sociology. The university of Chicago.
[13] Vakan L 2007*SocIs*. 6 93 –101 (Published in Russian)
[14] Zubok Yu A, Chuprov V I 2015 *Research Result. Sociology and Management series* **l(4(6))** 8 – 17 (Published in Russian)
[15] Wallerstein I 2000 *International Sociolog* **15** **(2)** 249–265
[16] Kachalov N, Kornienko A, Kvesko R, Nikitina Yu, Kvesko S, Bukharina Z 2015 *Procedia – Social and Behavioral Sciences* **(206)** 459–463
[17] Bleikher O, Kvesko S 2015 *Applied Mechanics and Materials* **770** 651 – 655
[18] Janszen F 2015 *The Age of Innovation (*London, Prentice Hall)
[19] Sosnin N V 2006 *Competence-based approach in innovative engineering education* (Krasnoyarsk: Krasnoyarsk. state. tech. un-t.) (Published in Russian)
[20] Yurchenko A V, Zotov L G, Mekhtiev A D, Yugai V V, Tatkeeva G G *2015 IOP Conf. Ser.: Mater. Sci. Eng.* **81** 012112
[21] Kvesko S, Kabanova N, Shamrova D (2016) *MATEC Web Conferences* **79** 01062 doi:10.1051/matecc onf/ 20167901062

Biometric identification

V I Syryamkim, D N Kuznetsov and A S Kuznetsova
Tomsk State University, Tomsk, Russia
E-mail: novorostomsk@gmail.com

Abstract. Image recognition is an information process implemented by some information converter (intelligent information channel, recognition system) having input and output. The input of the system is fed with information about the characteristics of the objects being presented. The output of the system displays information about which classes (generalized images) the recognized objects are assigned to. When creating and operating an automated system for pattern recognition, a number of problems are solved, while for different authors the formulations of these tasks, and the set itself, do not coincide, since it depends to a certain extent on the specific mathematical model on which this or that recognition system is based. This is the task of formalizing the domain, forming a training sample, learning the recognition system, reducing the dimensionality of space.

1. Introduction

The main operations in pattern recognition using the methods discussed are the operations of determining the similarity and difference of objects. The objects in this group of methods play the role of diagnostic precedents. At the same time, depending on the conditions of a particular task, the role of a separate precedent can vary within the widest limits: from the main and decisive to the very indirect participation in the recognition process. In turn, the conditions of the task may require the successful solution of the participation of a different number of diagnostic precedents: from one in each recognizable class to the full sample size, as well as different ways of calculating measures of similarity and distinction of objects. These requirements explain the further division of extensional methods into subclasses.

Table 1.3 contains a brief description of the methods of pattern recognition with respect to the following parameters:

- classification of recognition methods;
- scope of recognition methods;
- classification of limitations of recognition methods.

Table 1. Table of classification of methods of recognition, comparison of their areas application and limitations

Classification of methods recognition		Application area	Restrictions (limitations)
	Methods based on estimates of the distribution densities characteristic values (or similarities and differences of objects)	Problems with the known distribution, as a rule, normal, the need for a set of great statistics	Necessity of search the entire training sample in recognition, high sensitivity to the unrepresentation of the training sample and artifacts

	Methods based on assumptions about the class of decision functions	Classes must be well separated, a system of attributes - orthonormal	The form of the decisive function must be known in advance. The impossibility of taking into account new knowledge about correlations between signs
Intensive methods recognition	Logical methods	problems of small dimension feature spaces	When selecting logical decision rules (conjunctions) is a complete search. High computational complexity
	Linguistic (structural) methods	Tasks of small dimensions of the feature space	Recovery task (definition) of a grammar by a certain set statements (descriptions of objects), is it is difficult to formalize. Unresolved theoretical problems
Extensional methods recognition	Comparison with the prototype	Comparison with the prototype	High dependence the results classification from the measure of distance (metric). Uncertainty optimal metric
	Nearest Neighborhood Method	The method of the nearest neighbors The problem of small dimension by the number of classes and indications	High dependence of results classification from measure distance (metric). The need for a full enumeration of the training sampling in recognition. Computing labor intensity
	Evaluation algorithms (voting) of the ABO	Tasks of small dimension by the number of classes and indications	Dependence of classification results on the measure of distance (metrics). The need for a full enumeration of the training sample when recognition. High technical complexity
	Algorithms estimates (voting)	Tasks of small dimensionality by the number of classes and	Very high technical complexity method, the unsolved

		indications	nature of a number of theoretical problems, the definition of the areas of competence of private methods, and in the particular methods themselves

Neural network, fuzzy-multiple and hybrid models in systems for processing and analyzing multidimensional complex data, their advantages and disadvantages with different approaches to modeling complex systems. Structural and functional possibilities and limitations of individual classes of fuzzy models, neural networks and their compositions dominating under representation and analysis of the corresponding types of models of complex systems and processes.

Biometric systems consist of two parts: hardware and specialized software.

The hardware includes biometric scanners and terminals. They fix one or another biometric parameter (fingerprint, iris, veins on the palm or finger) and convert the information received into a digital model available to the computer. And the software processes this data, correlates it with the database and decides who was before the scanner.

In order for the biometric system to be able to further identify the user, it must first register information about its identifiers. Commercial systems do not store images of real identifiers, but their digital models. When the user repeatedly accesses the system, the model of his identifier is again formed, and it is compared with the models already entered earlier in the database.

Of the total biometrics market, 14.7% are occupied by face identification technologies. Identification by fingerprints incorporated the best that is inherent in biometrics in general. A fingerprint is identified by a fingerprint, unlike the password, the fingerprint can not be forgotten, transferred to another. Modern scanners have learned to establish the belonging of a fingerprint to a living person, and they can not be deceived, imprinting an imprint on paper, gelatin or glass. The probability of erroneous identification is 0.0000003%, and the time required for scanning the print does not exceed a fraction of a second [1–5].

2. Denial of passwords

Denial of passwords was made through the introduction of a fingerprint scanner into a smartphone. No matter how they scolded the Touch ID function, opponents of biometric data collection, the technology is actively used not only for simple unlocking of a smartphone, but also for making purchases or using services.

In the future, some large banks are also considering the possibility of switching to a fingerprint scanner as a way of authentication. Moreover, the companies are working on a prototype of the iris scanner, which will be used in the online payment service. Despite openness to the perception of a new one, buyers are more inferior to more invasive technologies, such as face recognition software, that could be used to identify them by an employee during a shopping trip.

Assessing fingerprint recognition technology in the process of paying for goods, almost half (42.4%) of respondents would welcome this technology if it also allowed them to automatically receive a home delivery service (Figure 1).

Figure 1. Fingerprint.

In addition, 57% of the buyers surveyed want to have the ability to scan the product from their devices to see reviews and recommendations for other products that they might like, while 48% of buyers do not object to the function of pop-up offers starting to arrive on their mobile device at the entrance to the store.

3. Cydercrime

On the black market, there are at least 14 vendors offering skimmers who can steal data from fingerprints, and at least three researchers who are working on technologies that can break systems for recognizing the pattern of veins on the wrist and iris. On the black market pre-sale testing of the first versions of biometric skimmers has already been carried out. Then several errors were found, but the main problem was the use of GSM modules for transmitting biometric data - they could not cope with large volumes of information, which means that new versions of such skimmers will use other, faster data transfer technologies [2]. In the communities of cybercriminals actively discussing the development of mobile applications that allow to mask human faces [2]. Such programs help to use photos of real people posted on social networks to deceive the face recognition system. Unlike passwords or PIN-codes, which can be easily changed in case of hacking, fingerprints or the image of the iris of the eye can not be changed. Accordingly, if biometric data is once in the hands of others, their continued use will pose a serious risk. That's why they need exceptionally reliable methods of protection. The danger lies also in the fact that they are introduced into modern electronic passports and visas, which means that the theft of such documents leads to the fact that in the hands of the attacker is virtually all information on which the identity of a person can be established. "

4. Voice biometry

Voice biometrics is one of the technologies that develops very quickly and allows different companies to use its solutions to identify customers. In the biometric system, individual behavioral, psychological and some other characteristics are used to determine or confirm the personality. There are many biometric measurements, including scanning the iris, fingerprints, face recognition, voice, signature, etc..

Voice biometry allows, by examining the voice characteristics of a person, to identify a person. It is a relatively simple and economical way to solve a number of practical problems. Voice biometry is a highly developed technology that can be used to improve the quality of a service to such an extent that a person can experience this improvement. The developer must provide the person with automated service, and speech technologies can help in this.

Nobody forces people to wait, does not redirect and does not offer to use the menu. Voice communications are convenient for a person.

The system understands the person and is able to test his words. He may not even remember the password or the number. The biometry of the voice, which is used during the conversation, allows you to determine who is calling. This reduces the talk time. So the client does not need to be presented and name the password. His password is his voice! At the same time, he feels that his call is important and the developers immediately make a decision [3–7].

The most widely used technology is applied in the banking sector, in insurance companies, in telecom. Airlines are showing considerable interest. The market of mobile applications for cell phones is also promising, where speech technologies are in full demand. In the automotive industry, voice systems allow you to use navigation devices along the way, are able to turn on music, air conditioning, help, without distracting from the control of the car, write down and send SMS, etc.

In medicine, speech technologies are used to record information about clients, create electronic patient maps. This allows you to optimize the work of doctors and creates clear benefits for customers. The doctor does not use the computer keyboard, he just dictates the medical indicators and diagnosis. The speech recognition system translates the voice into text and writes it down [8–15].

Bank contact centers successfully use voice technology. If the client needs basic information, then it is given to him freely. But if he wants to conduct a financial operation or some operation with his account, then his [status] should be checked. Voice biometry is one of the types of human verification, with which it is possible to identify whether a person is alive, or a speech record is broadcast (Figure 2).

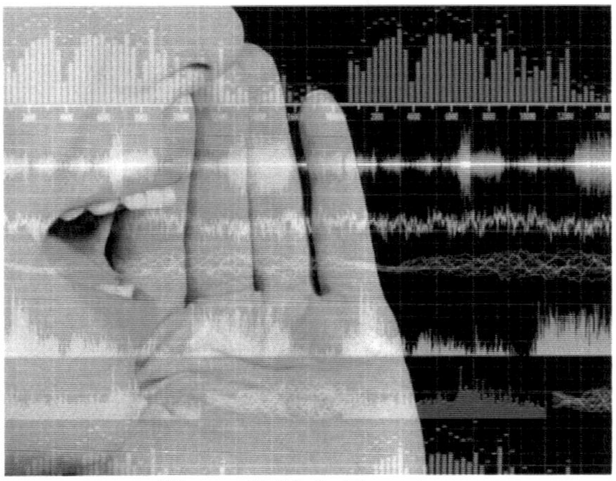

Figure 2. Vois biometry.

The voice biometry system can reveal the need for additional verification of a person. You can also create a `blacklist 'of client votes, seen in fraud or in attempts to unauthorized access to the accounts of other customers. This allows you to ensure the safety of banking operations.

5. Smell Detection

The ability to identify individuals using body odor as a method of unique human identification. Changing the smell can be evidence of a substitution.

The system, developed by the scientists of Madrid, is able to identify people by the smell emanating from the body. The researchers argue that the body of each person has permanent distinct "odor patterns", which are not affected by disease, diet, or age. Researchers have created a sensor that can recognize the "unique patterns" of the smells of the human body and identify their carrier with an accuracy of 87%. The sensor was tested on 11 volunteers, of which eight were men and five were women. Scientists took thirty samples of smell from cleanly washed hands of each of the subjects at different times of the day.

According to the developers, the sensitivity of the sensor was so high that it was difficult to deceive with soap, deodorant, cologne or other attempts to change the smell. In the official statement of the university, scientists express confidence that this opens the possibility for creating "less aggressive" ways of identifying a person than what exists at the present time. Despite the fact that the recognition of the iris and fingerprint gives high accuracy of identification, in the mass consciousness, these technologies are closely associated with forensic science, which causes distrust and protest, the scientists state [4,16].

Recognition of persons at the current stage of development gives too high a level of errors. Thus, the development of odor sensors, allowing to identify a person passing by them, opens the possibility for the development of more comfortable and inconspicuous methods of identification with a sufficiently high level of accuracy. Researchers are confident that such technologies can be used at airports, at checkpoints on the border and in any other situations where identification is currently applied on the photo. Identification by smell is one of the oldest methods used to find and identify people, but now forensic experts use specially trained dogs for this. The development of methods for effective recognition of the smell of man using electronic devices started relatively recently.

The most original and unusual method of identifying an individual in our opinion is to implant chips or swallow microcomputers. This will solve the problem of authorization radically and forever. Chips and microcomputers will analyze the pulse, the composition of the gastric juice and other internal biometric information. This, in turn, will ensure the security of Russia and the world.

Summarizing the results of this article, it can be argued that neural networks technologies for biometric authentication of users of open systems are technically feasible at the modern level of knowledge. Accordingly, for the creation of neural network technologies for biometric authentication, it is necessary to borrow the symmetrization of the problem of identifiable Volterra nuclei.

Acknowledgments
The article is prepared within the framework of Competitiveness Enhancement Programme of National Research Tomsk State University

References

[1] Yazov Yu K, Volchihin V I, Ivanov A I, Funtikov V A, Nazarov I G 2012 *Neural network protection of personal biometric data* (Moscow, Radio technics)

[2] Akhmetov B S, Ivanov A I, Funtikov V A, Bezyaev A V, Malygina E A 2014 *The technology of using large neural networks to convert fuzzy biometric data into the code of access key* (Almaty, LEM Publishing house)

[3] Bekhtin Yu S, Klestov S A, Kutsov M S, Syryamkin V I, Titov D V 2016 *Theoretical foundations of digital image processing in build-in vision systems* (Tomsk, STT Publishing house)

[4] Syryamkin V I 2017 *Intellectual robotic and mechatronic systems: textbook* (Tomsk, STT Publishing house)

[5] Kuznetsov D 2018 *Matec Web of Conferencec* **155** 01018 doi: 10.1051/matecconf/201815501018

[6] Syryamkin V I *et al.* 2009 *Devices and systems of automatic control of high accuracy. –* (Tomsk, Publishing house Tom. University)

[7] Syryamkin V I 2010 *Correlation-extreme radio navigation systems* (Tomsk, Publishing house Tom. University)

[8] Syryamkin V I *et al.* 2012 *Cognitive systems for monitoring and forecasting the scientific and technological development of the state* (Tomsk, Publishing house Tom. University)

[9] Syryamkin V I 2013 *Regional problems of public security* (Tomsk, Publishing house Tom. University)

[10] Syryamkin V I 2012 *Medical diagnostic and therapeutic systems. Principles of designing and building medical hardware and software systems* (Germany, LAP LAMBERT Academic Publishing)

[11] Syryamkin V I 2013 *Recognition of complex oil and gas deposits based on neuro-fuzzy portraits* (Germany, LAP LAMBERT Academic Publishing)

[12] Syryamkin V I, Bureev A Sh, Zhdanov D S 2009 A method for diagnosing the state of human or animal organs and the device for its implementation. Patent for invention № 2429979 from July 22, the Russian Federation

[13] Syryamkin V I 2014 *Neuro-fuzzy methods in intelligent systems for processing and analyzing multidimensional information* (Tomsk, Publishing house Tom. University)

[14] Titov V S *et al.* 2012 *Methods and systems for digital processing of aerospace images* (Novosibirsk, Science)

[15] Serikov M Z *et al.* 2017 *Proceedings – 2016 11th International Forum on Strategic Technology, IFOST 2016, 510*

[16] Yurchenko A V *et al.* 2015 *IOP Conf. Ser.: Mater. Sci. Eng.* **81** 012112

Mathematic modeling of the method of measurement relative dielectric permeability

I V Plotnikova, N V Chicherina and A B Stepanov
National Research Tomsk Polytechnic University, Tomsk, Russia
E-mail: inna@tpu.ru

Abstract. The method of measuring relative permittivity's and the position of the interface between layers of a liquid medium is considered in the article. An electric capacitor is a system consisting of two conductors that are separated by a dielectric layer. It is mathematically proven that at any given time it is possible to obtain the values of the relative permittivity in the layers of the liquid medium and to determine the level of the interface between the layers of the two-layer liquid. The estimation of measurement errors is made.

1. Introduction

The big variety of the measuring tasks connected with the necessity of carrying out the measurements of various liquid environments level [1] is characteristic for the various fields of modern science and technology [2, 3]. So, for example, the necessity of measurement of the level arises in the majority of intellectual systems of environmental monitoring and safety, in modern productions and also at control of weight and an expense of liquid environments at a stage of their transportation or in the process of storage [4, 5].

Now there are more than twenty various ways of measurement of liquid environments level [6–9]. A part of them was widely used in the modern industry, and the others have the highly specialized field of application because of various shortcomings inherent in them. As a rule, not only the level of the liquid environment [10], but also position of the section limit between layers [11, 12] is required not only to measure in the majority practical applications if the liquid environment is multicomponent. For example, oil / commercial water in settling tanks can be an example of such multicomponent liquid environment. It should be noted that the majority of modern practical approaches for measurement of the section limit of two-layer liquid are unsuitable. So, for example, a number of methods of measurements can work only if the controlled liquid environment is transparent [10]. Therefore, there is a problem of consideration of other technical solutions to the measurement of level and limit of the section of the two-layer liquid environment.

Now most for measurement of liquid environments level the capacitor sensors of level measurement are widely used [13–17]. The dependence of value of electric capacity of the measuring converter in the form of the condenser educated, for example, by means of two plates which are partially entered on the measured environment from liquid level height is the basis for the principle of operation of the capacitor level gage. In this article the measurement method which is based on the usage of three capacitor sensors of level is considered.

2. Research object and method

The considered method of measurement of relative dielectric permeability and position of the section limit between layers of the liquid environment describes the measuring system, consisting of three capacitor measuring converters in the form of flat condensers. It is necessary to notice that the electric condenser represents the system consisting of two conductors, which are divided by a dielectric layer. The value of electric capacity of the condenser C for a case of a plane-parallel arrangement of measuring electrodes can be determined by the following mathematical expression (at neglect by regional effect):

$$C = \varepsilon_0 \varepsilon \frac{s}{d}, \tag{1}$$

where ε is relative dielectric permeability of the interelectrode environment; ε_0 is a dielectric constant; s is area of measuring electrodes of the condenser; d is distance between measuring electrodes of the condenser.

The possible relative positioning and the geometrical sizes of measuring electrodes of three capacitor D1–D3 sensors are shown in the Figure 1. We will consider that the distance between measuring electrodes of sensors is equal d, and width and height measuring electrodes of capacitor sensors of level are equal a and b.

We will consider the tank with the two-layer liquid environment. We will consider at the same time that two-layer liquid occupies completely all volume of the tank. The relative dielectric permeability of the top layer of the liquid environment is equal ε_1, and relative dielectric permeability of the lower layer of the liquid environment ε_2 (Figure 2).

In this case the problem of measurement of position of the section limit of liquid environments comes down to use of three capacitor sensors of level. We will assume that the limit of the section of two liquid environments passes through measuring electrodes of the D2 sensor (Figure 2). We will consider that the beginning of coordinates is in the top boundary surface of the tank. We will consider at the same time that the D1 sensor is shipped in the controlled two-layer liquid environment on the H_1 level, the D2 sensor on the H_2, level, and the D3 sensor on the H_3 level, and the limit of the section of two liquid environments is at the level H_L (Figure 2).

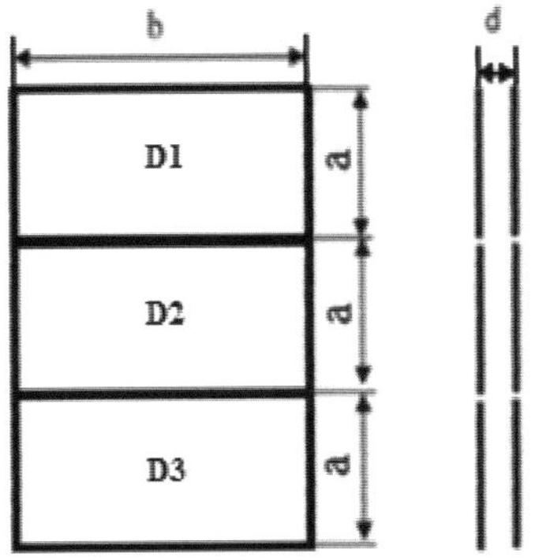

Figure 1. Arrangement and sizes sensors of level.

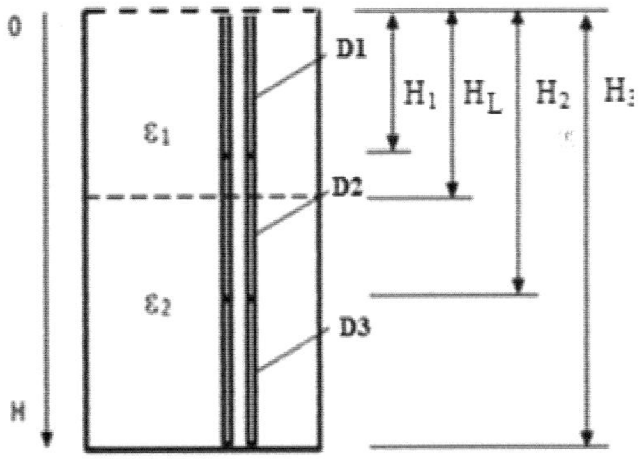

Figure 2. The tank with three capacitor sensors of level.

According to (1) we will find the value of capacity for each of three sensors of level.

We will write down value of capacity of the D1 sensor as:

$$C_1 = \frac{\varepsilon_0 \varepsilon b H_1}{d}. \tag{2}$$

We will write down value of capacity of the D2 sensor as:

$$C_2 = \frac{\varepsilon_0\varepsilon_1 b(H_L - H_1)}{d} + \frac{\varepsilon_0\varepsilon_2 b(H_2 - H_L)}{d}. \quad (3)$$

We will write down value of capacity of the D3 sensor as:

$$C_3 = \frac{\varepsilon_0\varepsilon_2 b(H_3 - H_2)}{d}. \quad (4)$$

From expressions (2) and (4) we will write down values of relative dielectric permeability as ε_1 and ε_2:

$$\varepsilon_1 = \frac{dC_1}{\varepsilon_0 bH_1}. \qquad \varepsilon_2 = \frac{dC_2}{\varepsilon_0 b(H_3 - H_2)}. \quad (5)$$

We will consider that $H_3 = 3H_1$, $H_2 = 2H_1$.

Having substituted the received expressions for ε_1 and ε_2 in expression (3), we receive the expression for value of section limit level of two liquid environments:

$$H = H_1 \frac{c_2 - 2c_3 + c_1}{c_1 - c_3}. \quad (6)$$

Thus, values of relative dielectric permeability of both layers of the liquid environment and the level of limit of the section between layers of two-layer liquid are known at any moment.

3. Assessment of errors of measurement

We will assume that in a system which is represented in the Figure 2 the measurement of capacity value happens with a margin error c_{ci} that is

$$c_1 = c_{1t} + \xi_{c1},\ c_2 = c_{2t} + \xi_{c2},\ c_3 = c_{3t} + \xi_{c3},$$

where c_{1t}, c_{2t} c_{3t} are true values of capacity of sensors of level, and ξ_{c1}, ξ_{c2}, ξ_{c3} are full errors of measurement of the corresponding capacities.

We will define errors of measurement of relative dielectric permeability ε_1, ε_2 and position of limit of sections H_L through errors of measurement of capacitor sensors of the level ξ_{C_1}, ξ_{C_2} и ξ_{C_3}..

At nonlinear dependence of position of limit of section H_L on the measured values of parameters according to (6) mean square deviation of result of indirect measurement is defined as follows:

$$\sigma_{\varepsilon 1} = \frac{d}{\varepsilon_0 bH_1}\sigma_{c1},\ \sigma_{\varepsilon 2} = \frac{d}{\varepsilon_0 b(H_1 - H_2)}\sigma_{c3}, \quad (8)$$

where σ_{c1}, σ_{c3} mean square deviations of capacitor D1 and D3 sensors.

At nonlinear dependence of position of limit of section H_L on the measured values of parameters according to (6) mean square deviation of result of indirect measurement is defined as follows:

$$\sigma_{H_L} = \sqrt{\left(\frac{\partial H_L}{\partial c_1}\right)^2 \cdot (\sigma_{c_1})^2 + \left(\frac{\partial H_L}{\partial c_2}\right)^2 \cdot (\sigma_{c_2})^2 + \left(\frac{\partial H_L}{\partial c_3}\right)^2 \cdot (\sigma_{c_3})^2} \quad (9)$$

At the same time

$$\frac{\partial H_L}{\partial c_1} = \frac{H_1(c_2 - c_3)}{(c_1 - c_3)^2}, \quad \frac{\partial H_L}{\partial c_2} = \frac{H_1}{c_1 - c_3}, \quad \frac{\partial H_L}{\partial c_3} = \frac{H_1(c_1 - c_2)}{(c_1 - c_3)^2},$$

Then $$\sigma_{H_L} = \sqrt{\left(\frac{H_1(c_2 - c_3)}{(c_1 - c_3)^2}\right)^2 \cdot (\sigma_{c_1})^2 + \left(\frac{H_1}{c_1 - c_3}\right)^2 \cdot (\sigma_{c_2})^2 + \left(\frac{H_1(c_1 - c_2)}{(c_1 - c_3)^2}\right)^2 \cdot (\sigma_{c_3})^2}$$

4. Modeling of work of a method with three capacitor sensors of level

The considered method of measurement of relative dielectric permeability and position of section limit between two layers of the liquid environment has been simulated in the MathCAD package of the following entry conditions:

- height of the tank is 0.6 m;
- height of measuring electrodes of capacitor sensors of the level is 0.2 m;
- width of measuring electrodes of capacitor sensors of the level is 0.3 m;
- distance between measuring electrodes of capacitor sensors of the level is 0.01 m;
- relative dielectric permeability of the first layer of the liquid environment is 2.2;
- relative dielectric permeability of the second layer of the liquid environment is 81;
- time of modeling is 1000 sec.

It was also supposed that errors of measurement of capacitor sensors of level represent white noise with an amplitude of 1% from extreme value of the indication of sensors.

In the course of modeling on every second selection of 10 measurements of values of capacities for each of three sensors of level was formed. Then the mathematical expectations and mean square deviations for values of capacities of sensors of level were determined by the received selection. Besides, on each measurement of selection were calculated relative dielectric permeability of liquid layers, and position of section limit between layers for which later the mathematical expectations and mean square deviations were defined.

In Figures 3 and 4 schedules with the results of mathematical modeling are submitted.

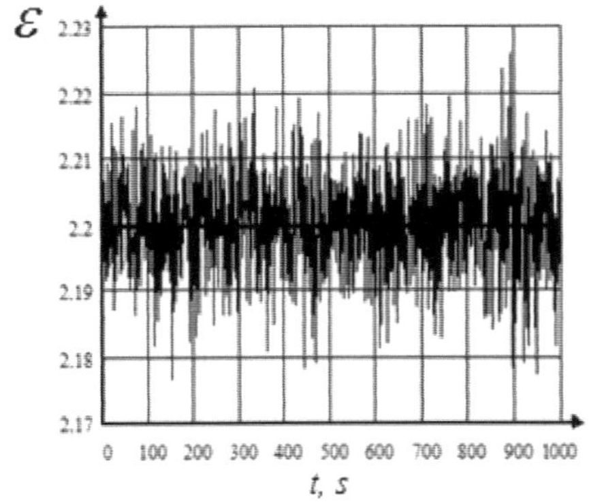

Figure 3. Relative dielectric permeability of the top layer.

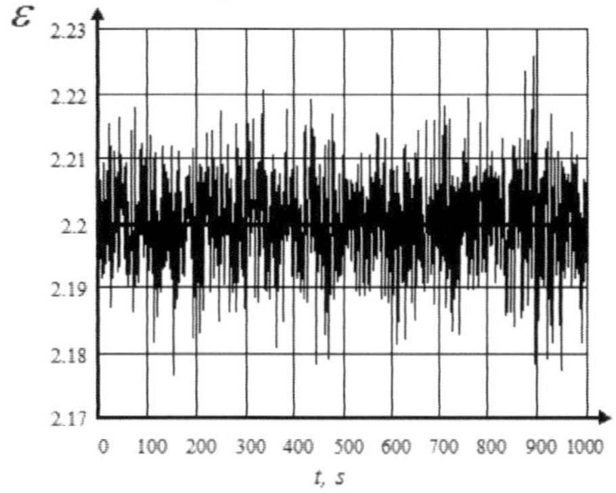

Figure 4. Relative dielectric permeability of the lower layer.

As the result of mathematical modeling we have received that for the capacitor D1 sensor the extreme value of capacity is 11.68 pF, and for the capacitor D3 sensor is 430.1 pF.

The value of amplitude of white noise for the D1 sensor will be $3\sigma_{c_1} = 0.12$ pF and the value of amplitude of white noise for the D3 sensor will be $3\sigma_{c_3} = 4.3$ pF.

Having substituted these found values in expressions (8), we will receive:

$$\sigma_{\varepsilon_1} = 2.26 \cdot 10^{-3}, \ \sigma_{\varepsilon_2} = 0.081.$$

As the expression (11) includes the current value of capacity of the D2 sensor, for determination of theoretical value σ_{Hs} will calculate it for timepoint of *t*=1000 sec.

In this time point C1=11.69 pF, C2=428.3 pF, C3=430.5 pF, H_L=0.20 m, and the mean square deviation for the position of section limit H_L will make $\sigma_{Hs} = 6.35 \cdot 10^{-4} m$.

5. Summary

In article the method of measurement of relative dielectric permeability of layers and position of section limit of the two-layer liquid environment based on the usage of three capacitor sensors of level is considered. The results of the mathematical modeling are given and assessment of errors of measurement is executed.

References

[1] Siggins A F 2011 *Measurement Sci. Technol* **22(2)** 1–9
[2] Gradoboev A V *and* Orlova K N 2014 *Adv Mater Res* **880** 237–241 doi: 10.4028/www.scientific.net/AMR.880.237
[3] Galtseva O V *et al* 2015 *IOP Conf Ser: Mater Sci Eng* **81(1)** 012062 doi: 10.1088/1757-899X/81/1/012062
[4] Nizhegorodov A I *et al* 2017 *IOP Conf Ser: Mater Sci Eng* **289** 012014 doi: 10.1088/1757-899X/289/1/012014
[5] Uglov V V *et al* 2013 *Russian Journal of Non-Ferrous Metals* **54(4)** 349-354 doi: 10.3103/S1067821213040159
[6] Wobschall D 1977 *IEEE Transactions on Geoscience Electronics* **15(1)** 49–58
[7] Korzenok I N *et al IOP Conf Ser: Mater Sci Eng* **110(1)** 012088 doi:10.1088/1757-899X/110/1/012088
[8] Wakamatsu H 1977 *Hewlett-Packard Journal* **48** 37–44
[9] Dexter A. R *et al* 2010 *Soil Science Journal* **175(7)** 320–328
[10] Petrova AY *et al* 2015 *Technical Physics* **60(4)** 592–594
[11] Gomez-Sanchez J A 2012 *Journal* of *Electrical Bioimpedance* **3** 29–35
[12] Nelson S O 2010 *J Microw Power Electromagn Energy* **44(2)** 98–113
[13] Ishida T *et al* 2000 *Clays and Clay Minerals* **48**(1) 75–84
[14] Wagner N *et al* 2010 *Proc of the 1st European Conference on Moisture Measurement, Aquametry* 228–237
[15] Surzhikov A P *et al* 2016 *IOP Conf Ser: Mater Sci Eng* **110(1)** 012002 doi: 10.1088/1757-899X/110/1/012002
[16] Wagner N *et al* 2011 *IEEE Trans. Geosci. Remote Sens.* **49(7)** 2518–2530
[17] Petrusev A S et al 2016 *2nd International Conference on Industrial Engineering, Applications and Manufacturing (ICIEAM 2016)* 7911461 doi: 10.1109/ICIEAM.2016.7911461

The capacitive proximity sensor based on transients in RC-circuits

A G Yakunin
Head of the Chair, AltSTU, Barnaul, Russia
E-mail: almpas@list.ru

Abstract. The principle of operation of the capacitive proximity sensor is described. It can be used in various robotic complexes, automation systems and alarm devices to inform the control device of the approach to the sensor sensitive surface of an object. At the heart of the device is the measurement of the change in the current of the transient accompanying the charge of the reference capacitor because of the parallel connection to it the capacitance formed by the sensitive sensor surface and the external object. At the heart of the device is the measurement of the change in the current of the transient accompanying the charge of the reference capacitor caused by the parallel connection to it the capacitance formed by the sensitive sensor surface and the external object. As shown by theoretical and experimental studies, the value of this capacity, depending on the purpose of the device, can vary within very wide limits. In this case, the sensitive surface can be both a piece of ordinary wire several centimeters long, and a metall plate or grid, the area of which can reach units and even tens of square meters. The main advantage of the proposed solution is a significant reduction in the effect of spurious leakage currents arising at the capacitance of the measuring electrode under the influence of pollution and humidity of the environment.

1. Introduction

Capacitive measuring converters of non-electrical quantities are now widely used along with other sensors in a wide variety of fields of science and technology, including automation of production processes [1–3] and in the field of robotics [4, 5]. With capacitive sensors, linear [6] and angular [7, 8] movements can be monitored, fluid level [9], tactile inforces [10, 11], pressure [12] and other physical quantities. It is also known to use capacitive devices as proximity sensors in production [13] and in alarm protection systems along with acoustic, radio wave and optical converters [14].

The main advantage of capacitive devices is their relative simplicity, small dimensions, low power consumption, manufacturability and low cost. Nevertheless, despite the above advantages, the field of capacitive converters using is rather limited. So, for example, as proximity sensors they are almost not used in chemical and dusty industries, as they change their characteristics in the presence of dust and under the influence of aggressive media. The inconstancy of the properties of the controlled environment leads to a change in the currents of parasitic leakage of the sensor capacitance, which causes a large number of false positives. Since for such applications of capacitive sensors the environment is the dielectric of the measured capacitance, and the surrounding things, controlled objects and surfaces are the electrodes, it is not possible to solve the problem by improving the technology of manufacturing the sensors themselves, as described, for example, in [4]. Therefore, for such applications, one of the promising ways to improve the technical and economic parameters of capacitive sensors is the use in them of new operating principles, the development of new circuitry solutions, as well as the improvement of methods for processing signals and extracting information from them, which is the subject of this work.

Therefore, for such applications, one of the promising ways to improve the technical and economic parameters of capacitive sensors is the use in them of new operating principles, the development of new circuitry solutions, as well as the improvement of methods for processing signals and isolating the information component from them, which is the subject of this work.

2. Analysis of existing decisions

At present, there are three main approaches for finding the capacitance value of a capacitive converter. The simplest is to measure the voltage at the output of the voltage divider, the circuit of which is shown in Figure 1*a* or is close to it. In this circuit, the generator G must generate a sinusoidal voltage. The disadvantage of this solution is the nonlinear dependence of the output signal on the capacitance value. In addition, this approach is least resistant to the influence of the leakage resistance not shown in the figure, which shunts the capacity of the converter. Because of this, at the present time the basis of most capacitive sensors is the circuit in which the unknown capacitance is the frequency-setting element of the generator (Figure 1*b*) and enters the parallel or series oscillatory circuit or in the frequency-dependent positive feedback loop of the multivibrator.

In order to improve the accuracy of the measurement and increase the sensitivity of the sensor, it is often use not direct measurement of the frequency but a variety of difference schemes, when the capacitance is determined from the beat frequency after the summation of the frequencies of the reference and measuring oscillator, or by changing the duty cycle of the measuring pulse relative to the reference pulse. But in all variants of such solutions it is completely impossible to get rid of the parasitic resistances influence. In addition, in all cases a nonlinear relationship exists between the measured capacitance and the informative parameter. To exclude the effect of parasitic resistance, it is possible in impedance meters, when, for example, in the circuit shown in Figure 1*a*, not only the amplitude but also the phase of the output voltage is measured. However, such solutions are significantly more complex and have not been used in such simple devices as proximity sensors.

Therefore, for those cases when parasitic resistance is unstable, varies widely, and determination of its value is not required, it is more promising to apply solutions when the capacitance is determined from the parameters of the transient processes, namely, by the value of the capacitor charging current at the beginning of the transient process after feeding to it voltage. One of the variants of such a solution is shown in Figure 1*c*, where Sm is the input preamplifier with function of summation and current-to-voltage converter, AU – automatic control unit, [15]. In such a device, the capacitance value is determined from the voltage value U_1, which is adjusted by the adjustment system to minimize the total current getting to the preamplifier input. If the voltage U_1 has the same sign as U_0, the control voltage to the key K_1 must be supplied via the inverter shown in the figure with a dashed line. The advantage of this solution is the linear dependence of the output signal on the value of the capacitance C_2 measured and its weak dependence on the leakage currents flowing between its plates Если напряжение имеет тот же знак, что и, управляющее напряжение на ключ K_1 должно подаваться через инвертор, показанный на рисунке пунктирной линией. This is explained by the fact that at the moment of switching the charge current of the capacitance has a maximum value and, in the absence of resistance of the connecting conductors, does not depend on the value of the resistance connected in parallel to the measured capacitance through which the leakage currents flow. However, in such a scheme it is not possible to completely get rid of the influence of spurious leakage resistance, since its decrease tightens the duration of the charge pulse of the capacitor and increases its amplitude. Therefore, as the leakage currents of the measured capacitance increase due to an increase in, for example, the humidity of the medium and contamination of the surface on which the measuring capacitor covers are placed, the compensating voltage U_1 will decrease, which is equivalent to decreasing the measured capacitance. In addition, in the diagram shown in Figure 1c, the measuring capacitor is not grounded, which significantly limits the possibilities of using this circuit in automation systems, since in them one of the covers of the measuring capacitor is most often an earthed object.

The modifications proposed in [16] of the scheme presented in Figure 1*c*, described in the next subsection and shown in Figure 2, are deprived of the indicated shortcomings.

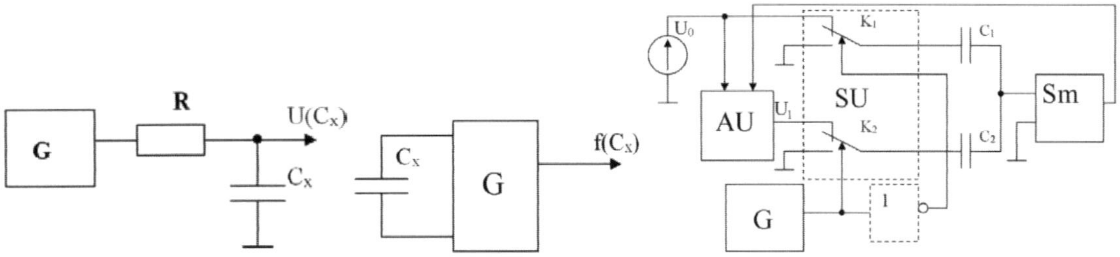

a) b) c)

Figure 1. Variants of information signal generation circuits in capacitive sensors.

3. The proposed approach to minimizing the effect of spurious leakage currents on the operation of a capacitive converter

The first fundamental difference of the solution proposed in [16] from the one shown in Figure 1*c* is that the second terminal of the measured capacitance C_X is not connected to the input of the current to voltage converter, but is grounded. This allowed to significantly expand the scope of possible application of the device for the above reason. In this case, an increase in the value of the capacitance measured in this circuit does not lead to an increase, but to a decrease in the overcharge current of the capacitor C_1, but the linear coupling between the capacitance and the current of the recharge current is kept in a sufficiently wide range.

The second difference is that at thc output of the converter a sampling-storage scheme is added, so that only a very short time interval is selected from the entire time diagram, corresponding to the start of the process of recharging the compensation and measuring capacitors. Due to this, even more than the prototype, the effect of the parasitic resistance of the measuring capacitor on the magnitude of the output signal was suppressed.

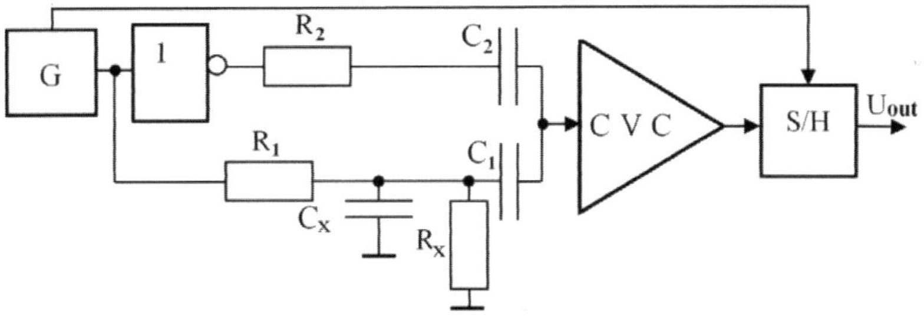

Figure 2. Structural diagram of the proposed solution: G – Rectangular pulse generator, CVC – Current to Voltage Converter, S/H – sample and hold circuit.

4. Results of calculations and experimental studies

For the analysis of the operation of the circuit shown in Figure 2, the model of its measuring cell, shown in Figure 3, was used. The transfer of one of the terminals of the capacitor C_1 from the preamplifier input to the ground is entirely acceptable, since in the current to voltage converters their input voltage is almost invariably and close to zero.

IOP Conf. Series: Materials Science and Engineering **363**(2018) 012007 doi:10.1088/1757-899X/363/1/012007

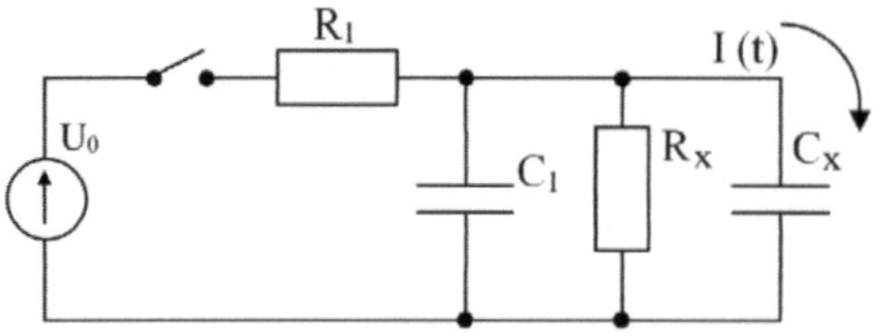

Figure 3. The equivalent circuit of a rechargeable cell with a measured C_X and a reference C_1 capacitor and a shunt resistor R_X through which a leakage current flows.

For such a circuit, the voltage across the capacitors and the current of their charge when the key is closed will be described by the expressions

$$U_c(t) = R_X \cdot U_0 \frac{1-\exp\left(-t\cdot\frac{R_1+R_X}{R_1\cdot R_X\cdot C}\right)}{R_1+R_X}; \quad (1)$$

$$I(t) = U_0 I(t) = U_0 \frac{C\cdot\exp\left(-t\cdot\frac{R_1+R_X}{R_1\cdot R_X\cdot C}\right)}{C\cdot R_1}, \quad (2)$$

where $C=C_1+C_X$.

In order to explicitly assess the effect of capacitance C_X and leakage resistance R_X on the values of currents and voltages described by these expressions at different instants of time, let us consider the corresponding sensitivity functions representing their C_X derivative:

$$F_U(t) = U_0 \cdot \frac{R_1+R_X}{R_X+(R_X\cdot C)^2} \cdot t \cdot \exp\left(-t \cdot \frac{R_1+R_X}{R_1\cdot R_X\cdot C}\right); \quad (3)$$

$$F_I(t) = -U_0 \cdot C_X \frac{R_1\cdot R_X\cdot C - t\cdot(R_1+R_X)}{R_X R_1{}^2 C^3} \cdot \exp\left(-t \cdot \frac{R_1+R_X}{R_1\cdot R_X\cdot C}\right). \quad (4)$$

The graphs of the voltage and current sensitivity versus time curves for different values of R_X for the case when R_1 = 5 kΩ; C_1 = 200 pF and U_0 = 10 V, are shown in Figure 4.

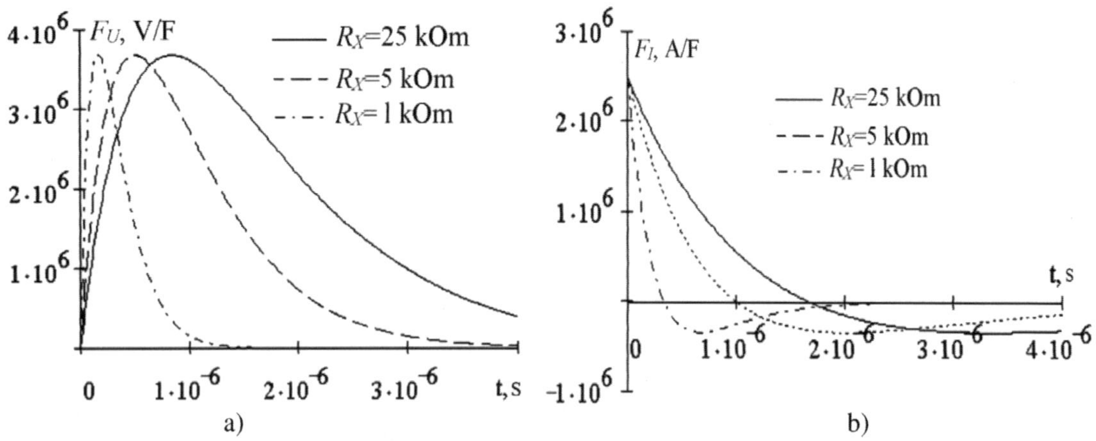

Figure 4. The dependence of the voltage sensitivity (*a*) on the capacitor C_X and the current flowing through it (*b*) on the time for different values of the leakage resistance R_X.

As can be seen from the expressions (3.4) for $F_U(t)$ and $F_I(t)$ and the corresponding graphs, in both cases the value of the resistance R_X does not affect the magnitude of the maximum sensitivity. However, for the function FU (t), the position of the maximum sensitivity peak depends on the value of R_X, which makes it very difficult to use the $U(t)$ dependence for determining C_X. If we use the dependence of the charge current on time to find the capacitance, then, as can be seen from equation (4) and the graph in Figure 4*b*, it is possible to get rid of the influence of R_X on the measurement results if the charge current is measured at the shortest time from the start of the key switching. With reference to the realizing method of measuring the C_X circuit, shown in Figure 2, this moment corresponds to the voltage drop at the output of the generator G. At this moment, the sensitivity value is maximal and equal to

$$F_I(0) = -U_0 \frac{C_X}{R_1(C_1+C_X)^2}. \tag{5}$$

As follows from this expression, the sensitivity value here is also not constant, depends on the ratio between the capacities of the capacitors C_X and C_I and assumes the maximum value at $C_X = C_I$. The form of dependence $f_I(C_X/C_I) = -R_I \cdot F_I(0)/U_0$ normalized on the value U_0/R_I and on C_I is shown in Figure 5.

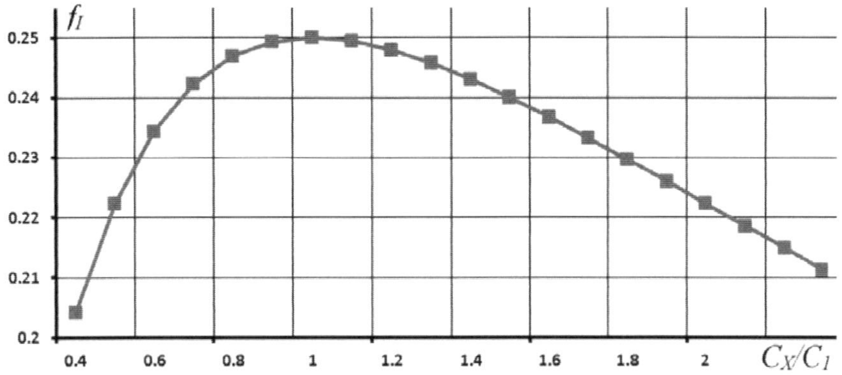

Figure 5. Dependence of the maximum of the relative sensitivity function on the current from the ratio of the measured capacitance to the capacitance of the reference capacitor C_I .

As can be seen from this dependence, when the measured capacitance deviates from the capacity of the reference capacitor by more than 2 times, the sensitivity decreases by approximately 20% of its maximum possible value. In this case, a decrease in the capacitance C_X relative to C_I is less preferable than its increase, since at $C_X \rightarrow 0$ the sensitivity drops to zero much more rapidly than when it increases. It follows that for best results it is necessary to select the values of the capacitances $C_I = C_2$ in accordance with the expected range of variation of the quantity C_X.

Along with the need to reconcile the circuit parameters with the value of the measured capacitance in order to achieve high performance properties of the sensor, it was also necessary to solve a number of other problems related to the optimization of the circuitry solutions used in it. For this purpose, a sensor model was developed that allows to take into account factors such as the spread of the parameters R_I, C_I, R_2, C_2 of the direct and compensating RC circuits, the dependence of these parameters on temperature, the presence of a delay on the inverting element, the finite duration of the pulse generator edges, the imperfection of the sample and hold circuit and current-to-voltage converter. Comparison of simulation results with experimental data, an example of which is shown in Figure 6, showed good adequacy of the developed model.

The optimization of the capacitive converter carried out with the help of this model made it possible to design control devices taking into account the specific conditions of their operation and functional purpose. Thus, a capacitive proximity sensor constructed using the developed optimization technique provided a sensitivity of 100 mV / pF at a sensitivity threshold of 0.1 pF in the range of measured capacitance from 50 to 500 pF, a temperature range from -40 to +45 ° C, and a parasitic shunt resistance from 5 kΩ to infinity. And the current consumption of such a sensor does not exceed 0.35 mA at a supply voltage of 5 V.

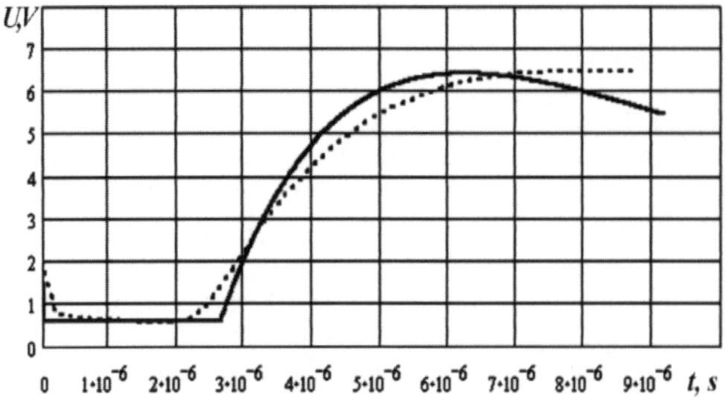

Figure 6. The signal at the output of the current to voltage converter at the beginning of the switching. The dashed line is the experimental result, the solid line is the results of mathematical modeling in the MathCAD environment.

5. Summary

The proposed method for constructing a capacitive converter gives a real improvement in reliability, sensitivity, and other technical and economic characteristics of proximity sensors developed on their basis. Parasitic leakage current in the dielectric, on which the measuring electrode of the converter is fixed, do not have practically no effect on the output signal, which is an undoubted advantage of this solution. But, since the output signal of the converter is nonlinearly dependent on the value of the capacitance being measured, it is advisable to use this converter primarily as an approximation sensor or another element of discrete automation used in various industries. In particular, it can be successfully used in the highly dusty conditions of production, connected, for example, with grain processing [17, 18], and also in the technical means of burglar alarm as a preemptive action detector [14]. In the latter case, to reduce the number of false positives and increase the reliability of reliable identification of unauthorized access attempts, one can use algorithms presented, for example, in [19, 20]. In addition, the application of the sensor together with modern means of microcontroller technology such as Arduino or Raspberry allows implementing promising methods of digital signal processing that can significantly increase the reliability of its operation. The use of microcontrollers or System-on-a-Chip will allow the use of such sensors in automatic control or non-destructive testing systems, provide the possibility of automatic tuning to a measurable capacitance, and also check the efficiency of the sensor and identify abnormal situations such as a break in the measuring electrode or its shorting to a ground.

Acknowledgments

The author expresses his deep gratitude to the leadership of the Altai State Technical University named after I.I. Polzunov (AltSTU) for financial support of the publication of this article and former graduate student AltSTU Avtsinov V.B. for participating in conducting theoretical studies, performing experimental studies of the proximity sensor, as well as for developing and implementation of equipment using the developed sensors

References

[1] Wilson J 2005 *Sensor Technology Handbook* (Elsevier)
[2] Dunn W 2006 *Introduction to Instrumentation, Sensors, and Process Control* (Norwood, Artech House)
[3] Clarence S 2007 *Sensors and Actuators: Control System Instrumentation* (Boca Raton,

CRC Press)

[4] Weadon T, Evans T, Sabolsky E 2013 Ceramic–polymer capacitive sensors for tactile/force awareness in harsh environment robotic applications Smart Mater. Struct. **22(12)** 125017 doi: 10.1088/0964-1726/22/12/125017

[5] Qiu S, Huang Y, He X et al. 2015 *Meas. Sci. Technol.* **26(10)** 105101 doi:10.1088/0957-0233/26/10/105101

[6] Picotto G, Pisani M, Sosso A 2009 *Meas. Sci. Technol.* **20(8)** 084011 doi: 10.1088/0957-0233/20/8/084011

[7] Hou B, Zhou B, Song M et al. 2016 *Sensors* **16(8)** 1196 doi:10.3390/s16081196

[8] Zheng D, Zhang S, Wang S et al. 2015 *IEEE Trans. Instrum. Meas.* **64(1)** 143–153 doi**:** 10.1109/TIM.2014.2328456

[9] Qurthobi A, Iskandar R, Krisnatal A and Weldzikarvina 2016 *J. Phys.: Conf. Ser.* **776** 012118 doi: 10.1088/1742-6596/776/1/012118

[10] Surapaneni R, Guo Q, Xie Y et al. 2013 *J. Micromech. Microeng.* **23(7)** 075004 doi: 10.1088/0960-1317/23/7/075004

[11] Lee H, Chung J, Chang S and Yoon E 2011 *J. Micromech. Microeng.* **21(3)** 035010 doi:10.1088/0960-1317/21/3/035010

[12] Narducci M, Yu-Chia L, Fang W and Tsai J 2013 *J. Micromech. Microeng.* **23(5)** 055007 doi: 10.1088/0960-1317/23/5/055007

[13] Båvall L, Karlsson N 1998 *Meas. Sci. Technol.* **9 (3)** 505–509 doi: 10.1088/0957-0233/9/3/027

[14] Suchkova L, Yakunin A 2017 *IOP Conf. Series: Materials Science and Engineering* **289** 012028 doi:10.1088/1757-899X/289/1/012028

[15] Grinevich F, Novik A 1987 *Measuring compensation and bridge devices with capacitive sensors* (Kiev, Science. Dumka)

[16] Yakunin A, Avtscinov V 2000 Capacitive converter *Patent RUS* 2168729

[17] Borisov A 2018 *IOP Conf. Ser.: Mater. Sci. Eng.* **289** 012039 doi: 10.1088/1757-899X/289/1/012039

[18] Nicolaeva B, Zlochevskiy V, Borisov A 2017 *Journal of Physics: Conference Series* **881** 012023 doi: 10.1088/1742-6596/881/1/012023

[19] Yakunin A, Hussein H 2017 *J.of Physics: Conf. Series* **881** 012028 doi :10.1088/1742-6596/881/1/012028

[20] Hussein H, Yakunin A 2015 *J. of Siberian Federal University Mathematics & Physics* **8(2)** 157–164

Cognitive methodology for forecasting oil and gas industry using pattern-based neural information technologies

O Gafurov[1], D Gafurov[2], V Syryamkin[1]
[1]National Research Tomsk State University, Tomsk, Russia
[2]Krasnoyarsk Research Institute of Geology and Raw Materials, Krasnoyarsk, Russia
E-mail: oleg.gafurov@mail.ru

Abstract. The paper analyses a field of computer science formed at the intersection of such areas of natural science as artificial intelligence, mathematical statistics, and database theory, which is referred to as "Data Mining" (discovery of knowledge in data). The theory of neural networks is applied along with classical methods of mathematical analysis and numerical simulation. The paper describes the technique protected by the patent of the Russian Federation for the invention "A Method for Determining Location of Production Wells during the Development of Hydrocarbon Fields" [1–3] and implemented using the geoinformation system *NeuroInformGeo*. There are no analogues in domestic and international practice. The paper gives an example of comparing the forecast of the oil reservoir quality made by the geophysicist interpreter using standard methods and the forecast of the oil reservoir quality made using this technology. The technical result achieved shows the increase of efficiency, effectiveness, and ecological compatibility of development of mineral deposits and discovery of a new oil deposit.

1 Introduction

The discovery-driven data mining technology presented in the article is based on the concept of patterns proposed the authors; this concept reflects many aspects of data relationships. Patterns represent regularities inherent in subsamples of data and they can be compactly expressed in a human-readable form.

In the *NeuroInformGeo* geoinformation system, the technique of the intellectual analysis of geological, geophysical and geochemical parameters based on the neural network methods with ranking of input attributes by the significance level is realized. The significance of the input characteristic provides an assessment of the informative contribution of this attribute to the result of the neural network training and realization of the forecast by the trained neural network.

The technology presented assumes the implementation of tuning and training of neural networks on the etalon field, which was brought into development with the appropriate geological complex and representative array of geological and geophysical data, which describe this etalon complex. The neural networks trained on the etalons form a pattern that can be used at the stages of exploration and additional exploration of oil and gas deposits in productive horizons of already discovered deposits. Taking into account the statistical data, they also can be used at the search stage on unexplored and poorly explored areas.

2 Pattern of the geological and geophysical complex

Accumulation of the experience in forecasting and interpretation can be ensured in the form of formation, preservation, and correct application of patterns that represent the library of trained neural networks on etalon deposits or drilled areas [4].

2.1 Pattern determination

A pattern is the accumulated experience of forecasting in the form of a library of trained neural networks on etalon objects.

Regarding to geological prospecting, reference objects are deposits or drilled areas that is objects with such a level of exploration, which allows checking the results of the neural network training using the reliable data on the geological structure of the etalon object.

2.2 Formation of the pattern of the geological and geophysical complex

The study of the entire volumetric part of the geo-environment is performed by geophysical methods (seismic and magnetic prospecting, etc.). During the process of natural fields studying (magnetometry, gravimetry, etc.) or based on compulsorily excited fields (seismic prospecting and electrical exploration), a three-dimensional matrix is formed where real geological objects are replaced by digital images of their physical fields.

Let us introduce the concept of a physical model of the medium in the form of a function $Q_n(x, y, z)$, which describes the distribution of a physical parameter in *A* region, which creates a geophysical field $f_n (x, y, z)$ in the surrounding space with geomedia parameters $p_1, . p_k$, and also the relation:

$$f_n (x, y, z) \Rightarrow f(Q_n, P_1,.., P_k), \qquad (1)$$

It connects Q_n and f_n, that is allows calculating the field of the given geophysical parameter (the direct problem of geophysics).

To calculate the field of the geophysical parameter f_n (x, y, z), we need the values of $Q_n (x, y, z)$ and parameters of the geo-environment $p_1, ., p_k$, as well as the position of the boundary (contact) surfaces. In seismic prospecting, this includes the value of the rock density and the boundary of their differences; in geology, this includes the value of clay and sandstone density, i.e. the density value of geological bodies and objects.

The problem of constructing a geological model $Q_n (x, y, z)$ using the geophysical parameter $f_n(x, y, z)$ measured is called the inverse problem. It is necessary to find distribution of parameters of the geo-environment $p_1 \ldots p_k$ and position of the contact surfaces.

Let us imagine that formation of answers for each step is a statement of *P (1), P (2), P (3), ...*, each of which can be either true or false, i.e. it may bring us to the problem solution or lead away. The principle of finite induction states that in order to prove the truth of propositions for all P (n) it is sufficient to establish the truth of P (1) and the truth of an infinite sequence of implications:

$$P(1) \Rightarrow P(2) \Rightarrow P(3) \Rightarrow \ldots P(n) \Rightarrow P(n+1) \Rightarrow \ldots \qquad (2)$$

The machine must correctly calculate the parameters at every step; and the initial problem must be set appropriately to a real geological environment. On each cycle, a neurocomputer program calculates the parameters based on the training sample.

2.3 An example of creating a pattern based on the etalon deposit in East Siberia

At the first stage of forming trained neural networks (patterns), an etalon deposit is determined, which was brought into development (Figure 1). Therefore, it has a representative array of geological and geophysical information, which will allow correct training and testing of neural networks and creating a pattern for further forecast of field and geological parameters in nearby promising areas (licensed areas - LA) poorly explored by exploratory drilling.

When forming a pattern in *NeuroInformGeo* geoinformation system [5], all input parameters (maps (meshes) of seismic attributes, geophysical, geochemical and other parameters) are normalized to the range [-1...1]. Before feeding to the synapse, each input signal is recalculated according to the equation:

$$Y_i = 2\times(x_i - min_i) / (max_i - min_i) - 1 \quad (3)$$

where x is the initial signal, Y_i is the received normalized signal, *min* and *max* are respectively the minimum and maximum values of the input parameters interval in the field fed to the synapse *i*.

During the neural network training, the data studies of wells are used, in which the tests and a complex of geophysical studies (GIS) were performed. A multilayer neural network is trained according to the backpropagation algorithm (*backward propagation of errors*). The network architecture is selected empirically providing a minimum error.

For each of the input parameters, the significance indicators are determined, which characterize the informative contribution of this parameter (attribute) to the results. When solving the *q*-o example, the significance indicator of the parameter is determined by the following formula:

$$\chi^q_{N_a} = \left| \frac{\partial H_q}{\partial w_{N_a}} (w_{N_a} - w^*_{N_a}) \right| \quad (4)$$

where, when solving the *q*-o example, the significance indicator shows how much the value of the estimation function changes in case of solving the q-o example by the network if the current value of the parameter w_{N_a} is replaced by the nearest selected value $w^*_{N_a}$ for the N_a parameter. The final significance indicator of the N_a parameter is calculated as the total average:

$$\chi_{N_a} = \frac{1}{n} \sum_{q=1}^{n} \chi^q_{N_a}, \quad (5)$$

where n is the number of examples. Thus, the calculated value of the significance indicator for the N_a parameter essentially represents (in the linear approximation) the absolute value of the change in the estimation function when the input parameter (signal) is removed from the network. Non-informative parameters are removed from the network, the neural network is retrained.

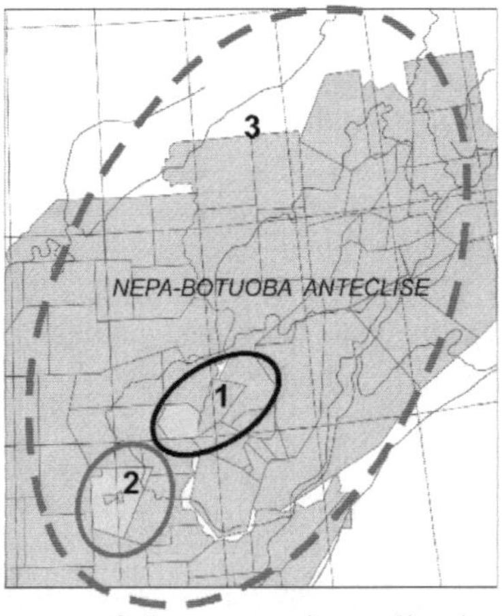

Figure 1. Zones of a possible forecast of the reservoir quality based on the Pattern created using seismic data and neuro-information technologies. The first zone means formation of etalon neural networks for productive horizons within the geological complex studied (Pattern). The second zone is a forecast of the reservoir quality (effective thickness, productivity, and permeability) on poorly explored licensed area. The third zone means the zone where it is possible to forecast the quality of the collectors based on the etalon pattern created.

The process of multiple iterations results in calculation of a true mean error of the neural network generalization.

If the training results are unsatisfactory, the initial geological and geophysical data are rechecked and the neural network is reset based on another feature that also characterizes the environment under the study.

If the solution to the problem is satisfactory, a pattern is formed, which is further used to forecast geological and field parameters on poorly explored LAs associated with an etalon deposit (based on which the pattern was created) by general geological and genetic characteristics.

Forecasting is carried out according to the etalon set of parameters corresponding to the pattern. The program classifies the multidimensional feature space into classes of membership (similarity) to a certain etalon and generates a map of etalons for the area interpretation or forms a cut with distinguishing complexes (classes) as temporary or deep models.

Figure 2 shows one of the patterns created. It characterizes the established dependence of the values of the effective thicknesses of Osinsky productive horizon (Nepa arch) on geophysical parameters (a complex of attributes of the seismic wave field), which describe the geological environment very accurately thus allowing training the neural network to forecast characteristics of the productive layer quality.

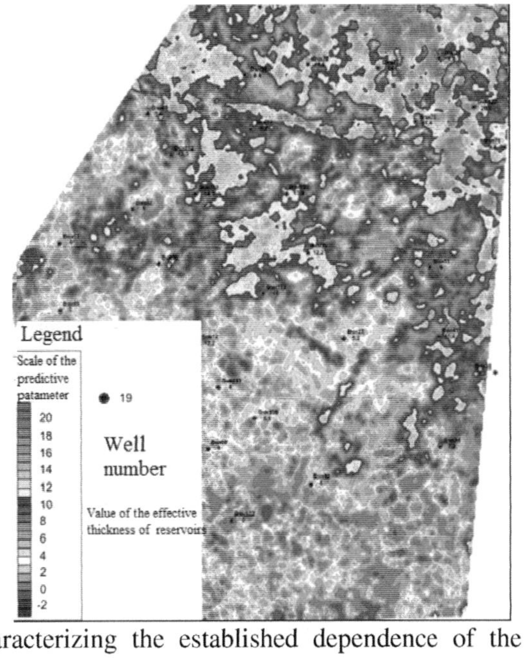

Figure 2. The pattern characterizing the established dependence of the values of the effective thicknesses of Osinsky productive horizon (Nepa arch) on the complex of attributes of the seismic wave field.

Training of the neural network was conducted in the borehole environment of exploratory wells according to a set of seismic attributes. In addition to the effective thicknesses *(N_{ef})* of the reservoir, the predicted parameters also included the porosity coefficient *(K_{por})*, the permeability coefficient υ, and various inflow characteristics. One of the most obvious classifications for inflows is carried out based on hydrodynamic studies conducted in exploratory wells (HDSW) that characterize the inflow type (lack of inflow, water, gas, oil). However, this classification, like the flow rate classifier, largely depends on the quality of drilling and tests conducted in different periods of the deposit exploration. For each predicted parameter, a neural network with layered architecture was created; it included up to 8–10 layers of 50–100 neurons in each hidden layer.

The training was conducted using the conjugate gradient method. Training using the conjugate gradient method showed the fastest convergence in estimating minimization of the error function (with a given accuracy).

During the training to predict the N_{ef} parameter, the neural network learned to recognize (predict) the values in the wells in 985 training cycles; besides, with a given accuracy of 1 meter, the average error was 0.6 meters, the maximum error was 0.9 meters, and all examples were solved correctly. When configuring a neural network to recognize K_{por} in the well area, all the examples included in the training sample were also solved correctly with a given accuracy of 1%. During the classification according to the HDSW data, there were no contradictory examples in the training sample and all examples were solved.

Testing of the trained neural network by cross-validation on test wells makes a big error. For example, when predicting N_{ef}, the average forecast error is 2.9 m, and the maximum error is 6.4 m (Table 1).

Table 1. Results of testing of the etalon neural network to forecast the N_{ef} parameter.

Well number	**N_{ef} according to GIS (m)**	**Average value of the forecast N_{ef} (m) R=50 m**
67	0	2.7
27	5.2	6.7
51	6	6.3
103	7.2	7.6
59	8.8	12.4
72	9.2	11.4
75	11.4	11.2
122	20.8	15.5
40	4	3.8
Average error:	**2.9**	
Max error:	**6.4**	
Max scatter:	**2.8**	

However, this range of error values is acceptable, taking into account that the forecasting is performed using a complex of seismic attributes, the vertical calculation range of which reaches 0.03 s. and more, which is because of the oscillation period length. Accordingly, the resolving power of the forecast depends on the wave properties of the time section and the method of measuring seismic characteristics and is determined primarily by the seismic parameter analysis range. At the same time, for some attributes, the calculating range exceeds the capacity of the Bilirsky Series and hence the productive horizon, which is of interest to us, because it covers thickness of not only a productive horizon, but also overlying impermeable layers of rock salt.

Thus, in a complex interpretation, the informative contribution of these attributes is based on an integral approach. The physical meaning of the integral analysis of dynamic attributes of the seismic wave field is presence of the seismic field in section of anomalous objects. The models of environments characterized by local unconsolidated areas, stress state of rocks, hydrocarbon dispersion, tectonic disturbances, and reef formations meet the requirements for forming anomalies [6,7].

Figure 3 shows the dependence of the predicted N_{ef} values of the effective thicknesses of the reservoir on the N_{ef} values in the wells according to GIS (numbers of test wells are signed). The dependence is linear, the square of the linear regression coefficient is 0.78, taking into account all the values predicted by the neural network in a certain confidence radius around the well. If considering only the average predicted N_{ef} values, the value of the square of the linear regression coefficient increases and amounts to 0.84 (correlation coefficient 0.92) on the average (for different sets of training and testing samples), which is excessive accuracy for solving seismic data prediction problems and sufficient for the pattern created. In this case, the maximum error of 6.4 m appears in the wells with the maximum effective thickness values determined according to the GIS data.

Thus, the neural network slightly understates the maximum values of the effective thickness in the wells with the maximum N_{ef} values. However, in general, it correctly divides the wells in quality predicting zones with no reservoirs as well as predicting the N_{ef} values (in meters) in the test wells and other parameters characterizing the reservoir quality.

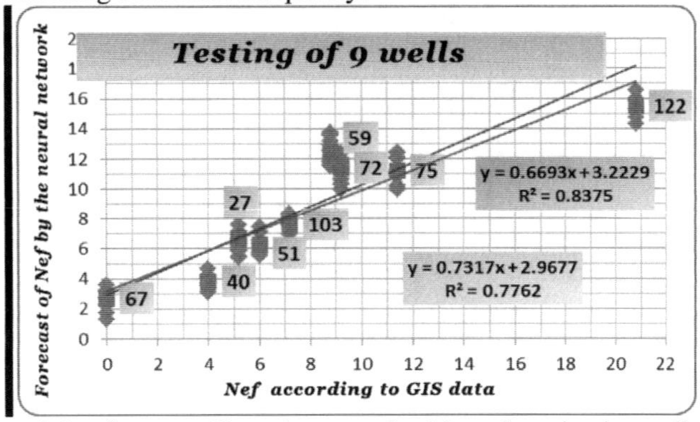

Figure 3. Dependence of the forecast N_{ef} values on the N_{ef} values in the wells according to the GIS data.

In order to understand the physical nature of the possibility of creating a pattern and forecasting the quality of reservoirs, we determined the attributes making the most significant contribution to the training and forecast of each parameter.

In general, when analyzing the forecast parameters, coefficients of the attribute significance were determined (as a result of the neural networks training in different productive horizons). Energy of scattered waves, acoustic impedance, and frequency and amplitude characteristics of the seismic wave field make a significant contribution to the forecast of effective thicknesses.

Scattered waves occur at supercritical angles of reflection from diffracters located in rock mass. Fractures, cracks, and caverns can be such diffracters. Acoustic impedance is a product of velocity in a layer and density. Thus, the change in impedance is associated either with a change in the propagation velocity of longitudinal waves, or with a change in density, and this relationship is linear. Besides, the velocity decreases with increasing porosity, increases with increasing pressure, decreases with increasing temperature, and decreases with fluid saturation.

For perspective strata, during the process of determining the reservoir quality, the lateral variation of the dominant frequency F1 and the amplitude of the value of the scattered component in the analysis range also play a significant role in the forecast. Lateral changes in the dominant frequencies F1, F2, F3 may indicate a fluid saturation or increased fracturing, which may occur because of changes in stratigraphy, lithology, and fault tectonics. As an example, Figure 4 shows the position of the dominant frequencies on the amplitude-frequency spectrum. The attribute of instantaneous frequency and instantaneous amplitude is obtained because of the Hilbert transformation of the original seismic trace. However, the physical meaning of the attributes after the Hilbert transformation does not change.

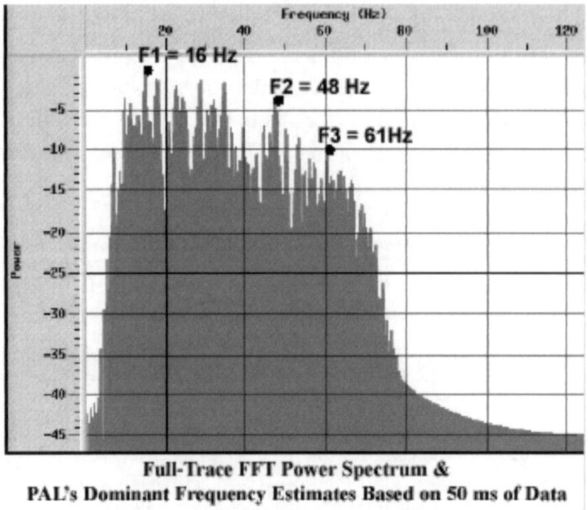

Figure 4. Position of the dominant frequencies on the amplitude-frequency spectrum.

For promising strata, the effective width of the amplitude-frequency spectrum also makes a significant contribution; the change of this parameter depends on the change in the dominant frequencies on the amplitude-frequency spectrum.

Amplitude parameters make a significant contribution to the forecast for all layers. The change in the amplitudes of the seismic traces corresponds to a change in the reflection coefficients. It follows from the Zoeppritz equation that under normal incidence of the P-wave on the boundary, no exchange waves appear, and the reflection coefficient is determined by the known impedance relation. However, in the case of a non-normal incidence, the Poisson's ratio plays an important role, or simply the ratio of the longitudinal velocity to the transverse velocity. In general, a change in the amplitude of reflected waves may indicate a change in lithology, density, fluid saturation, porosity, fracturing, and others.

The anomalies of the most significant attributes are confined to the layers of the positive structure in the context of the standard 3D cube, which may reflect the nature of the coastal-marine sedimentation conditions. The determined anomalies of the 3D cube coincide with the most significant attributes of the focusing transformation established during the training on the N_{ef} parameter - the maximum positive and mean square amplitudes of the scattered component of the seismic field. In the physical sense, this attribute reflects the most probable zones of decompression of the medium expressed in the increase of fracture and voidage. Attributes of the focusing transformation [8] also make a great contribution to the forecast of porosity and inflow classes.

In general, creation of the pattern resulted in forming the neural networks, which were trained to forecast the following parameters on the territory of the Nepa arch:

- effective thickness of layers;
- porosity coefficient;
- the nature of saturation (absence, water, oil, gas);
- effective pore volume;
- coefficient of permeability

The pattern was formed based on the data of exploratory wells (more than 100 wells) and 3D seismic data of the studied deposit of the Nepa arch in Irkutsk region (Russia).

2.4 Application of the Pattern on the licensed areas

On the licensed areas (LA) neighboring to the etalon deposit (Pattern) and located within the boundaries of Nepa arch in Irkutsk region, the same complex of geological and geophysical data (as on the etalon deposit) was calculated. Dynamic processing of the seismic data was carried out during the formation of data for forecasting improved reservoir properties along the Osinsky,

Preobrazhensky, and Ust-Kutsky horizons of some LAs. Attributes of the seismic wave field were calculated including AVO, acoustic impedance, and scattered-wave energy. A trend model of impedances was constructed; the results of GIS data interpretation and documents on wells testing in productive horizons were prepared. These activities resulted in forecasting and construction of forecast maps for a number of reservoir parameters. Figure 5 gives an example of comparison between the N_{ef} forecast on one of the LAs using a standard technique for interpreting geological and geophysical data and interpretation based on the neural networks of the pattern created. The forecast of the northern anomaly was confirmed later by a well drilling; the deposit was discovered.

Thus, the methodology developed which is oriented on combining the principles of training and interpretability corresponds to the goals and objectives of Data Mining. It is based on the concept of patterns that reflect fragments of multidimensional relationships in data.

These patterns represent the regularities inherent in the subsamples of data, they are expressed in a human-understandable form. The search for templates was carried out by the methods not limited to a priori assumptions about the structure of the sample and the form of distribution of the values of indicators analyzed.

In the geoinformation system *NeuroInformGeo,* a digital geological and geophysical model was obtained for this etalon deposit and a trained neural network, which preserves the coupling coefficients between geological and field parameters and attributes of the measured geophysical fields.

This model contains digital data and an analysis tool that allows adaptive training and retraining of neural networks (based on new data or simulation of the projected wells) as well as constructing geological and geophysical models taking into account the models built at the previous stage.

3 Conclusion

The realized approach within the framework of Data Mining (discovery-driven data mining), expansion of the model and its adaptive training provides wide opportunities to refine the forecast of reservoir quality on the areas studied and poorly studied, which in turn makes it possible to clarify the financial and economic assessment of new wells, etc. Creation of new etalons (patterns) will make it possible to generalize geological and geophysical information on other territories, establish regularities affecting the efficiency of geological and exploration work, and increase efficiency of geological and exploration work.

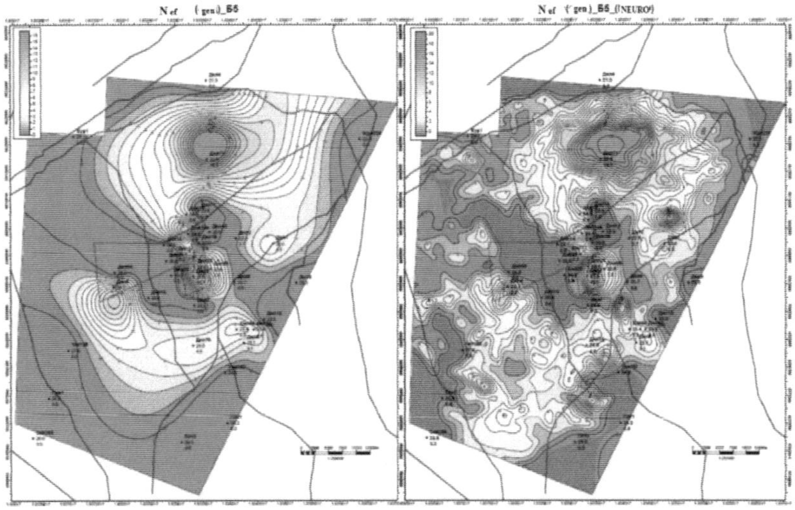

Figure 5. Comparison between the standard methodology for interpretation of geological and geophysical data and interpretation based on the neural networks of the pattern created.

Acknowledgments
This study (research grant No № 8.2.24.2018) was supported by the Tomsk State University competitiveness improvement programme. The authors are grateful to Tatiana B. Rumyantseva from Tomsk State University for her assistance in preparing the paper.

References
[1] Gafurov O M, Gafurov D O, Gafurov A O, Pankov M V, Bitner A K, Krasilnikova N B, 2012 Patent of the Russian Federation for invention No. 2477499 "A Method for Determining Location of Production Wells During the Development of Hydrocarbon Fields", application No. 20111125437.
[2] Gafurov O M, Gafurov D O, Gafurov A O 2011 Certificate of the Federal Service for Intellectual Property for the state registration of the computer program "The Neural Network-Based Program for Processing Seismic Acoustic and Geophysical Data in .dat Grids Format Used for Formation of Etalons and Calculation of Patterns", No. 2011613919
[3] Gafurov O M, Gafurov D O, Gafurov A O 2011 Certificate of the Federal Service for Intellectual Property for the state registration of the database: "The Database for the Block of Automated Formation of Etalons in the Intellectual Geoinformation System "NeuroInformGeo" No 201162031
[4] Gafurov D, Gafurov O 2016 *MATEC Web of Conferences* **79** 01086 doi: 10.1051/matecconf/20167901086
[5] Gafurov D and Gafurov O 2016 *MATEC Web of Conferences* **79** 01055 doi: 10.1051/matecconf/20167901055
[6] Nikitin A A, Zemtsova D P, Dolinin A N 2003 *Geophysics* **(6)** 3 – 8
[7] Hunt J 1982 *Geochemistry and Geology of Oil and Gas* (Moscow, Peace Publ.)
[8] Pozdnyakov V A 2005 *Geology, Geophysics and Development of Oil and Gas Fields* **(7)** 30–36

Two modular neuro-fuzzy system for mobile robot navigation

M V Bobyr [1], V S Titov [1], S A Kulabukhov [1], V I Syryamkin [2]
[1]Department of Computer Engineering, Southwest State University, Kursk, Russia
[2]National Research Tomsk State University, Tomsk, Russia
E-mail: max_bobyr@mail.ru

Abstract. The article considers the fuzzy model for navigation of a mobile robot operating in two modes. In the first mode the mobile robot moves along a line. In the second mode, the mobile robot looks for an target in unknown space. Structural and schematic circuit of four-wheels mobile robot are presented in the article. The article describes the movement of a mobile robot based on two modular neuro-fuzzy system. The algorithm of neuro-fuzzy inference used in two modular control system for movement of a mobile robot is given in the article. The experimental model of the mobile robot and the simulation of the neuro-fuzzy algorithm used for its control are presented in the article.

1. Introduction

The Moving of the mobile robot on a flat surface with the unknown location of obstacles complicate the development of neuro-fuzzy systems for controlling mobile robots. It is connected with the lack of information about the obstacles on the mobile robot's movement from the starting point to the finish. In this connection developers of the control systems for mobile robots' navigation use complex software codes and algorithms for determining the contours of objects in their way.

In their paper Zhu and Yang [1], Lee and Chiu [2] described the control system of the navigation for mobile robot based on neuro-fuzzy method. Later it was improved by Algabri and etc. [3] due to the introduction into it soft computing and genetic algorithms. Pandey and etc. [4] in their work presented a model for bypassing obstacles based on adaptive neuro-fuzzy inference system (ANFIS) model. The output signal in this model is the angle of rotation of the mobile robot. However, in the article is not specified how output linear function for the ANFIS is formed. Besides it is not indicated how the signal for turning of the mobile robot is transmitted from the microcontroller to its engines. When operating with help of MotorShield driver it is only possible to transmit the signals of PWM and delay time to the port of engine control. Exactly the angle turning of the mobile robot depends on delay time.

Traditionally, for controlling of the mobile robot turning used encoders. It complicates hardware and software implementation of the mobile robot, as well as increasing of its power consumption.

Limitation of the review works is the fact that the rigidly mounted on platform of the mobile robot ultrasonic distance sensor does not allow to scan the area around the robot. Therefore, to scan the space around the mobile robot and determining distances to obstacles it is necessary to rotate it around itself. Installation of ultrasonic sensor on the spindle servo will allow to scan the space around the mobile robot without rotation itself.

Thus, the problem of a mobile robot control and synthesis of methods of formation of the signals received from the sensors of the mobile robot and transmission them from the microcontroller to its engines is actual.

To solve this problem the authors, propose to use the system of decision-making control based on neuro-fuzzy inference. The mobile robot hardware is implemented on the microcontroller Arduino MEGA.

2. Kinematic scheme of the mobile robot

2.1. Structural scheme of the mobile robot

The proposed mobile robot uses a four-wheeled platform, two digital line sensors, three infrared distance sensors, ultrasonic distance sensor located on the spindle servo and microcontroller Arduino Mega (Figure 1).

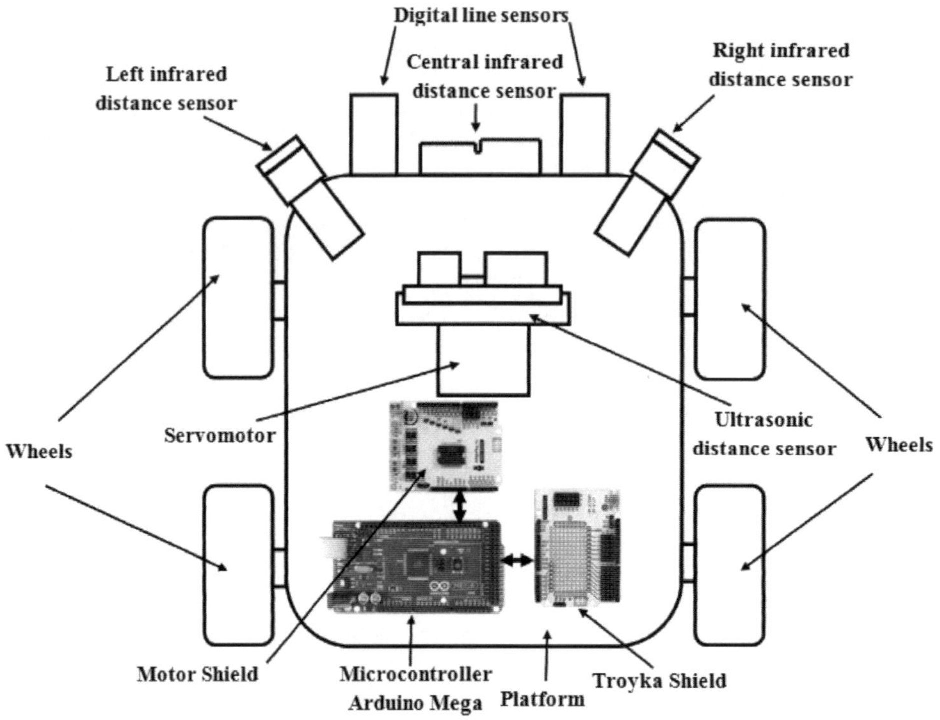

Figure 1. Mobile robot.

Connecting the engines to the Arduino MEGA board performed through the Motor Shield driver. Motor Shield is installed on top of the Arduino MEGA board. Four distance sensors and two sensors lines are connected to the Troyka Shield board. Troyka Shield is installed on top of the Motor Shield board. Detailed connection scheme Motor Shield and Troyka Shield drivers is presented in Figure 2.

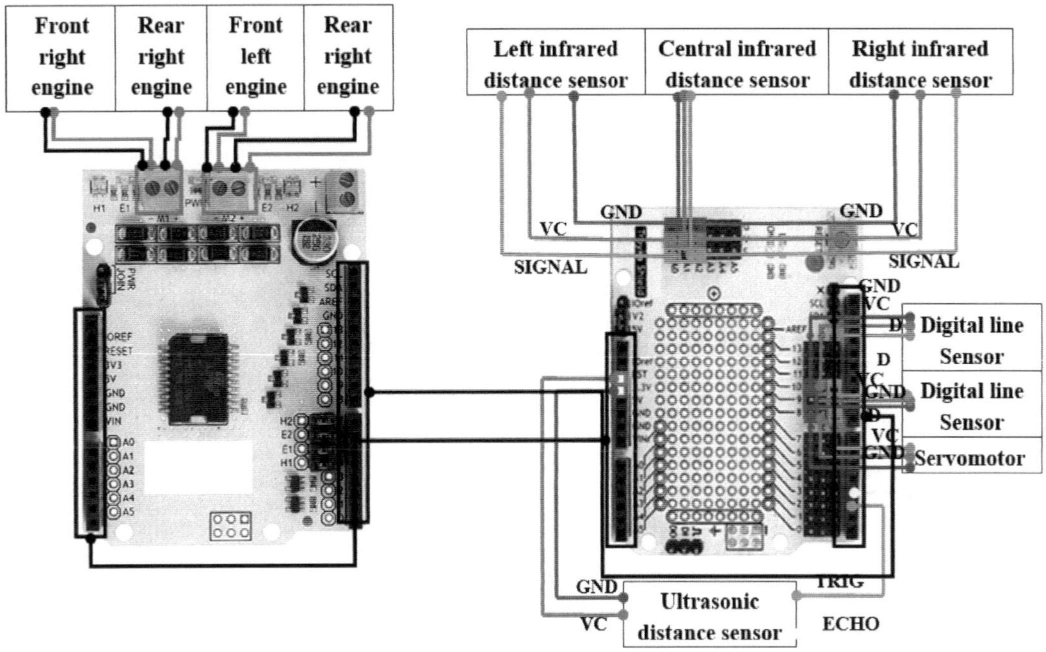

Figure 2. Connection scheme Motor Shield and Troyka Shield.

2.2. Statement of the problem for mobile robot movement

Let the mobile robot operate in two modes. In the first mode the mobile robot moves along the line. Two digital line sensors are used to control its movement, located in the front part of the robot, and fixed below the platform (Figure 1).

The second mode is to search for self-purpose and construction of the route of movement with the ability to bypass obstacles. Three infrared and one ultrasonic distance sensors are used to monitor the distance between the objects and the mobile robot. It is considered that the robot found the target when running over the surface indicated with a black cross. Thus, digital line sensors must be in logical unit state.

3. Modes of operation of the mobile robot

3.1. The first mode of operation of the mobile robot

In the first mode, when the mobile robot starts, power is passed to its motors and after that the robot moves along the line. There are two options for controlling the movement of the mobile robot along the line. In the first option used only one-line sensor which is fixed exactly above the line and the robot moves strictly along it. In the second option, two-line sensors are used, which are fixed on the both sides of the black line. The sensors monitor the white background, and if the robot runs over the black line, then they will switch to a logic zero. The authors in the work used a second method of controlling the position of the black line (Figure 3). The program code in which the moment of leaving the mobile robot from the black line is monitored, is shown in Figure 4 [5].

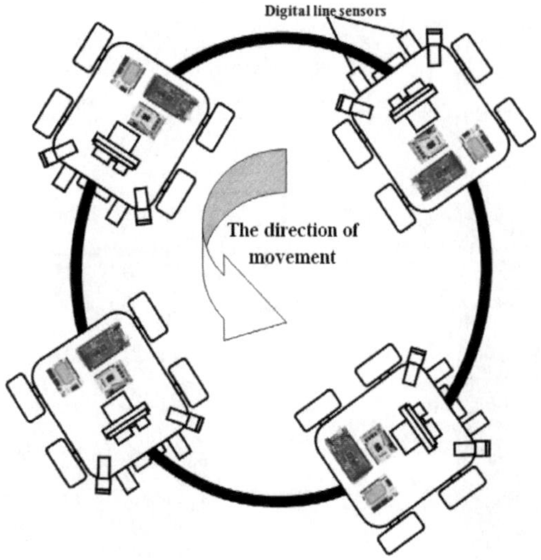

Figure 3. The first method of mobile robot movement.

```
int leftLinePin = 9;
int rightLinePin = 8;

int leftDirPin = 4;
int leftSpeedPin = 5;

int rightDirPin = 7;
int rightSpeedPin = 6;

int runSpeed = 125;

void setupMotorShield(){
  pinMode(leftDirPin, OUTPUT);
  pinMode(leftSpeedPin, OUTPUT);
  pinMode(rightDirPin, OUTPUT);
  pinMode(rightSpeedPin, OUTPUT);
}

void go(){
  analogWrite(leftSpeedPin, runSpeed);
  analogWrite(rightSpeedPin, runSpeed);
}

void stop(){
  analogWrite(leftSpeedPin, 0);
  analogWrite(rightSpeedPin, 0);
}

void goForward(){
  digitalWrite(leftDirPin, HIGH);
  digitalWrite(rightDirPin, HIGH);
  go();
}

void turnLeft(){
  digitalWrite(leftDirPin, LOW);
  digitalWrite(rightDirPin, HIGH);
  go();
}
void turnRight(){
  digitalWrite(leftDirPin, HIGH);
  digitalWrite(rightDirPin, LOW);
  go();
}

void setup (){ setupMotorShield(); }

void loop()
{
  boolean whiteLeft = digitalRead(leftLinePin);
  boolean whiteRight = digitalRead(rightLinePin);
  if (!whiteLeft && !whiteRight) { goForward();}
  else if (whiteLeft && whiteRight){ stop(); }
  else if (whiteRight){ turnLeft(); }
  else if (whiteLeft){ turnRight(); }
}
```

Figure 4. The program code of the robot movement along the line.

3.2. The second mode of operation of the mobile robot

In the second mode, the mobile robot independently performs the search of the way. Here two modular situations are possible.

The first modular situation. The mobile robot's movement along a straight line. To find the way it uses from the central infrared sensor the information about the obstacle located in front of it (Figure 5).

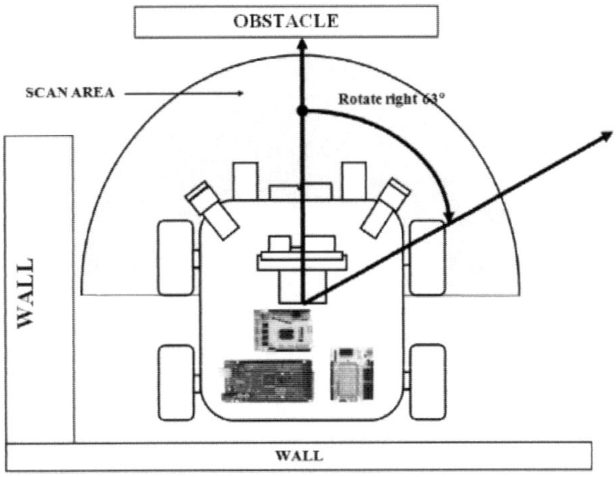

Figure 5. Searching an obstacle in front of the robot.

If the robot has found an obstacle, it stops. Next, the microcontroller passes commands to rotate the servo. Rigidly fixed the ultrasonic sensor on the spindle servo scans the area around the mobile robot (Figure 5, yellow background). After that is determined the distance to the most distant object and the angle to which it is necessary to rotate the robot for further movement. Next is calculation of the time delay for turning of the mobile robot to the resulting angle using fuzzy MISO-system [6–8]. Inputs of fuzzy MISO-systems are the speed of movement of the mobile robot and the angle of its turning. And the output is time delay required for turning of the mobile robot to the calculated angle (α=63°, Figure 5). After calculating the delay time, the robot is turning to the calculated angle and continues its movement.

The second modular situation. The robot moves along a narrow corridor and receives the information about an obstacle on left (L_1), in front (L_2) and on right (L_3) from the infrared sensors (Figure 6).

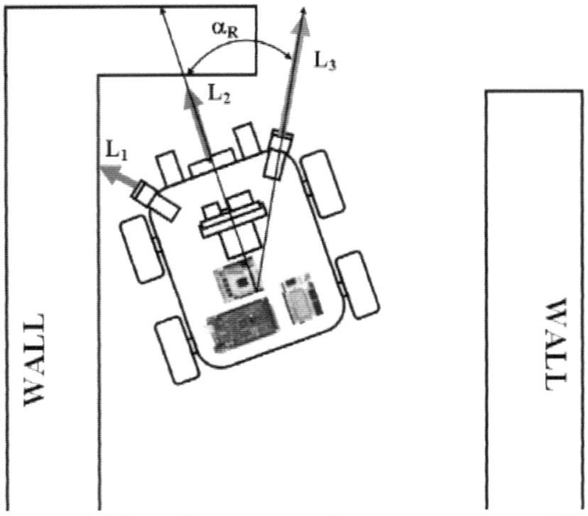

Figure 6. The robot moves along a narrow corridor.

If from the right infrared sensor (L_3) (Figure 6) received information about the absence of obstacles, then is used neuro-fuzzy system in which the result variable is the angle of the turning to the right (α_R).

The inputs of neuro-fuzzy system is the information about the distance to the object on the left (L_1) and in front (L_2). In the reverse situation, if we know the information L_2 and L_3 about the distance to the object then the neuro-fuzzy system will allow to determine the angle of the turning to the left (α_L). Considering the mentioned above, will give a block scheme of neuro-fuzzy MIMO-system (Figure 7).

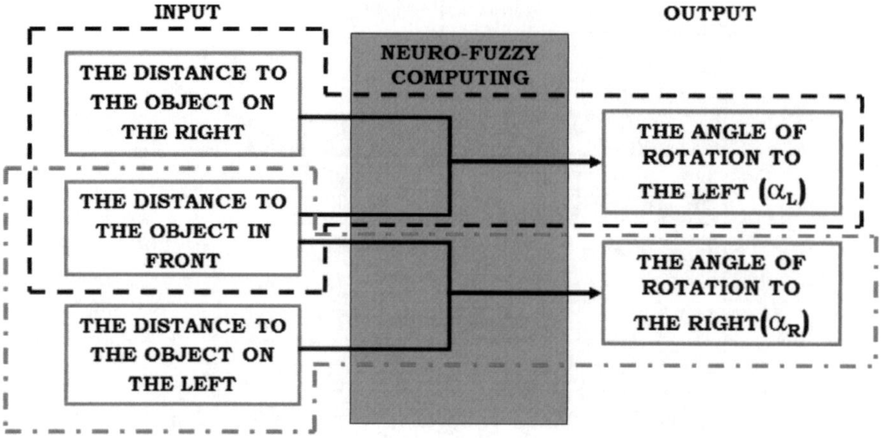

Figure 7. Neuro-fuzzy MIMO-system.

Conditions selecting of calculation output values of neuro-fuzzy MIMO-system depends on the rules (1) and (2).

$$\text{if } \max(L_1; L_2) < L_3 \text{ then } R, \quad (1)$$

$$\text{if } \max(L_2; L_3) < L_1 \text{ then } L. \quad (2)$$

On the arrival of the robot to a deadlock all the infrared sensors report about a short distance to the object. In order to get out of the deadlock, the robot moves backward and makes a 180° turn (Figure 8).

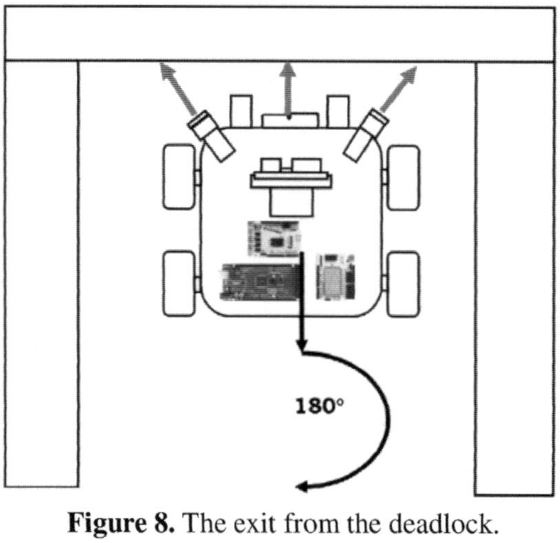

Figure 8. The exit from the deadlock.

The search of target is the main task of the robot when it moves in the second module situation (Figure 9). The completion of movement of the robot is the time when he moves up on the black cross. This moment is fixed by digital line sensors. After that the microcontroller generates a signal to stop the mobile robot.

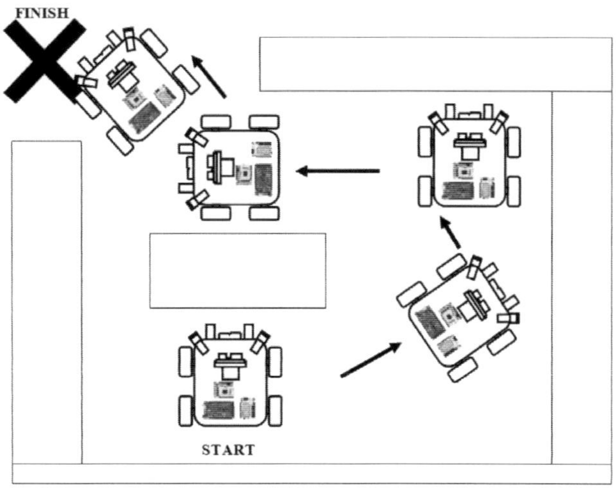

Figure 9. Target search.

4. Neuro-fuzzy control system for the mobile robot

To calculate the resulting variables of two modular the neuro-fuzzy MISO-system α, α_L, α_R is used the model described in the article [9–13]. The model consists of six sequential execution steps.

Step 1. Fuzzyfication of the input parameters. Triangular membership function is described by the formula (3), and the trapezoidal membership function - by formula (4).

$$f(x;a,b,c)=\begin{cases}0, & x\le a;\\ \frac{x-a}{b-a}, & a\le x\le b;\\ \frac{c-x}{c-b}, & b\le x\le c;\\ 0, & c\le x.\end{cases} \quad (3)$$

$$f(x;a,b,c,d)=\begin{cases}0, & x\le a;\\ \frac{x-a}{b-a}, & a\le x\le b;\\ 1, & b\le x\le c;\\ \frac{d-x}{d-c}, & c\le x\le d;\\ 0, & d\le x.\end{cases}, \quad (4)$$

where *a, b, c, d* are the parameters of the membership function; *x* is a quantitative value of the input parameter.

Step 2. Determination of values of MF for every implication of the input parameters, based on the information received from distance sensors.

Step 3. Synthesis of the knowledge base, which contains fuzzy rules of the type «If … then».

Step 4. Creation of a fuzzy ratio matrix. Soft arithmetic operations [14, 15] are used while calculating MIN (5) and MAX (6)

$$\min_{\delta}(x_1,x_2)_I=\frac{x_1+x_2+\delta^2+\sqrt{(x_1-x_2)^2+\delta^2}}{2}. \quad (5)$$

$$soft-\max(x_1,x_2)=\left|\gamma\cdot\max(x_1,x_2)+0{,}5(1-\gamma)(x_1+x_2)\right|. \quad (6)$$

Step 5. Defuzzification of the output parameters. This method is based on determining the area of triangular and / or trapezoidal membership functions [10] of input and output variables using a universal formula (7) (Figure 10).

$$S = \frac{h}{6}(b_1 + 4b_2 + b_3), \quad (7)$$

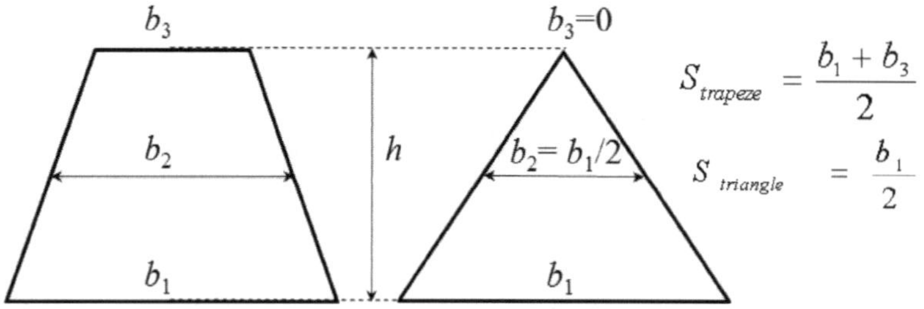

Figure 10. Calculation of geometric figures' area.

where h is height of the geometrical figure; b_1, b_2, b_3 are the width of low, middle and upper base of the geometric figure.

Using this method eliminates the need for global aggregation operations used in traditional fuzzy models.

Step 5.1. Determine the total area of the shape output value by formula (7).

Step 5.2. Determine areas of every output terms depending on the height values of the degrees of applicability.

Step 5.3. Determine areas of every output terms depending on the height values of the degrees of applicability by formula (8).

$$S_n = \begin{cases} S_n = 0, & than\ h = 0; \\ S_n = \frac{b_1}{2}, & than\ h = 1; \\ S_n = \frac{h}{2}(b_1 + b_3), & than\ h \in (0, 1). \end{cases} \quad (8)$$

Step 5.4. Determine the total area of the shape with transformed membership functions by formula (9).

$$S_{\text{trunc}} = \sum_{i=1}^{n} S_i. \quad (9)$$

Step 5.5. Determine the ratio of the total area of the shape of the output variable and the total area of the shape with transformed membership functions by formula (10).

$$D = \frac{S_{tranc}}{S_{glob}}. \quad (10)$$

Step 5.6. Determine the proportional relationship by formula (11)

$$y_{defuz} = \lfloor D \cdot (y_{final} - y_{start}) \rfloor + y_{start}, \quad (11)$$

where y_{final}, y_{start} are final and start values of the range of values of the output variables.

Step 6. The learning procedure. In this step, using the method of back propagation error occurs ANFIS learning. Here is a correction of transformed membership functions as long as the value y_{defuz} (11) becomes as close to the reference value according to the formula (12)

$$y_{out} = (w)_i + \delta(y_{defuz} - y_{\text{ref}}), \quad (12)$$

where δ is neuro-fuzzy inference system learning step (δ=0,01).

As shown in the article [10], the weighting parameter $\boldsymbol{w}$ is the ratio of $\boldsymbol{n}/2$ used to determine the total area. In the first step of learning where $\boldsymbol{w} = \boldsymbol{n} / 2 = 2.5$.

5. Experimental research

The scheme of the experimental model of the mobile robot working in two modes is shown in Figure 11. The mobile robot is collected on Pirate 4WD platform.

To test the efficiency of the proposed neuro-fuzzy model of mobile robot control system was performed simulation. To solve this problem, the program in Visual Basic language was written and implemented in MS Office Excel.

Figure 11. The experimental mobile robot.

The membership function for one of the input variables (L_1) has a triangular form (Figure 12).

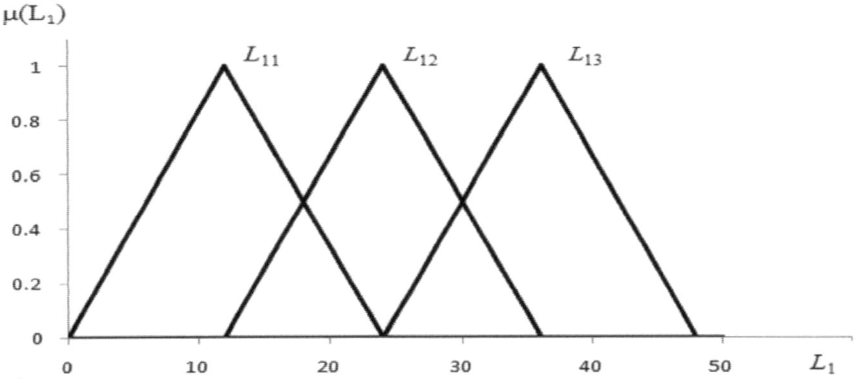

Figure 12. Triangular membership functions for the input variable - distance L_1

For example, when $L_1 = 5$, $L_2 = 8$, the result is being formed for output variable equal to α_R = 71°16' degree on output of neuro-fuzzy system with usage of hard formulas for searching minimum and maximum. When used in fuzzy inference soft arithmetic operations, this value is equal to α_R = 68°85'. The graph of the resulting surface is shown in Figure 13.

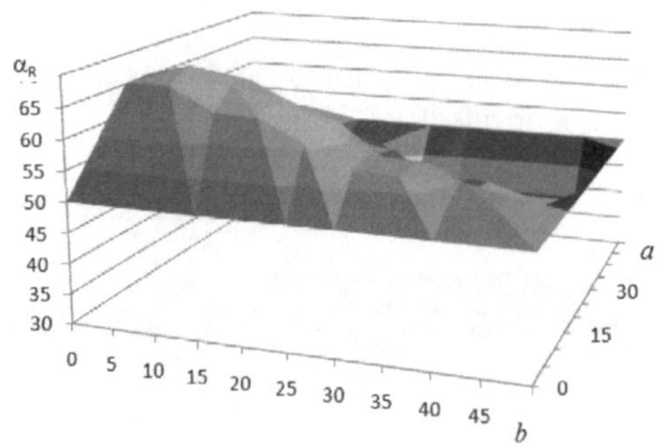

Figure 13. Resulting surface.

6. Conclusion

Using the soft arithmetic operations in the fuzzy inference [11–15] provides additive neuro-fuzzy models. It was established that defuzzification on the basis of the differential areas allows to take account of all information features arriving at the input of the neuro-fuzzy model. Experimental research of mobile robot functioning in two modes using neuro-fuzzy control system ensured the successful completion of all the tasks put in front of it.

Acknowledgments

This study (research grant No № 8.2.24.2018) was supported by the Tomsk State University competitiveness improvement programme, grant President RF MD-707.2017.8 and state assignment: Agreement № 2.3440.2017/4.6.

References

[1] Zhu A, Yang S X 2007 *IEEE Trans. Syst. ManCybern. Appl. Rev.* **37 (4)** 610—621

[2] Lee C H, Chiu M H 2009 *Expert Syst.Appl.* **36(5)** 8993—8999

[3] Algabri M, Mathkour H, Ramdane H, Alsulaiman M 2015 *CHB* **50** 42—56

[4] Pandey A, Kumar S, Pandey K K, Parhi D 2016 *Perspectives in Science* **8** 421–423

[5] Bobyr M, Luneva M, Yakushev A 2017 *MATEC Web of Conferences* **129** doi: 10.1051/matecconf/201712901064

[6] Bobyr M V, Yakushev A S and Milostnaya N A 2016. Fuzzy algorithm of a mobile robot's motion. International Conference on Industrial Engineering, Applications and Manufacturing. IEEE Inc. (DOI: 10.1109/ICIEAM.2016.7910971)

[7] Bobyr M V, Kulabukhov S A, Milostnaya N A 2016 *2ND International conference on industrial engineering, applications and manufacturing (ICIEAM)* doi: 10.1109/ICIEAM.2016.7910970

[8] Kruglova T N 2010 *Vestnik komp'yuternykh i informatsionnykh tekhnologij* **(8)** 28–35

[9] Bobyr M V, Milostnaya N A, Kulabuhov S A 2017 *Applied Soft Computing Journal* **59** 19–32 doi: 10.1016/j.asoc.2017.05.040

[10] Bobyr M, Titov V, Belyaev A 2016 *MATEC Web of Conferences* **79** 01052 doi: 10.1051/matecconf/20167901052

[11] Garcia-Rodriguez R, Parra-Vega V 2017 *Engineering Applications of Artificial Intelligence* **65** 43–50 doi: 10.1016/j.engappai.2017.07.013

[12] Sedov V A, Sedova N A, Glushkov S V 2016 *Vibroengineering PROCEDIA* **8** 453–458

[13] Sedova N A., Sedov V A., Glushkov S V 2016 *Vibroengineering PROCEDIA* **8** 506–511

[14] Bobyr M V, Kulabukhov S A 2017 *Journal of Machinery Manufacture and Reliability* **46(3)** doi: 10.3103/S1052618817030049

[15] Pham D T, Fahmy A A 2005 *In IFAC Proceedings* **16** 170–175 doi: 10.3182/20050703-6-CZ-1902.01453

IOP Conf. Series: Materials Science and Engineering **363** (2018) 012010 doi:10.1088/1757-899X/363/1/012010

LEDs based upon AlGaInP heterostructures with multiple quantum wells: comparison of fast neutrons and gamma-quanta irradiation

A V Gradoboev[1], K N Orlova[2] and A V Simonova[1]
[1]National Research Tomsk Polytechnic University, Tomsk, Russia
[2]Associate Professor, Department of Life Safety, Ecology and Physical Education, Yurga Institute of Technology, TPU Affiliate, Yurga, Russia
E-mail: gradoboev1@mail.ru

Abstract. The paper presents the research results of watt and volt characteristics of LEDs based upon AlGaInP heterostructures with multiple quantum wells in the active region. The research is completed for LEDs (emission wavelengths 624 nm and 590 nm) under irradiation by fast neutron and gamma-quanta in passive powering mode. Watt-voltage characteristics in the average and high electron injection areas are described as a power function of the operating voltage. It has been revealed that the LEDs transition from average electron injection area to high electron injection area occurs by overcoming the transition area. It disappears as it get closer to the limit result of the irradiation LEDs that is low electron injection mode in the entire supply voltage range. It has been established that the gamma radiation facilitates initial defects restructuring only 42% compared to 100% when irradiation is performed by fast neutrons. Ratio between measured on the boundary between low and average electron injection areas current value and the contribution magnitude of the first stage LEDs emissive power reducing is established. It is allows to predict LEDs resistance to irradiation by fast neutrons and gamma rays.

1. Introduction

AlGaInP heterostructures with multiple quantum wells are widely used as base of light-emitting diodes (LEDs) [1–7]. The LEDs are increasingly used in electronic devices for various applications. Under operating conditions they may be influenced by different radiation factors. Therefore, determining the interest to their resistance study as introduction of radiation defects leads to deterioration of the basic LED parameters [8].

AlGaInP heterostructures with multiple quantum wells are produced by metalorganic chemical vapor deposition (MOCVD)-based gaseous epitaxy [9–15]. However, a set of technological problems in production of heterostructures has not been solved yet. These are the following: hydrogen passivation, timely oxygen removal from the case, decrease of imperfection on heteroboundaries of layers and improvement of grown layer homogeneity.

These technological problems can cause high concentration of compounds (in particular Mg-H, Mg–O and Te-O) in the layers of AlGaInP heterostructures with multiple quantum wells [10], which are the centers of non-radiative recombination and/or absorption of active region-generated radiation. Aluminum in AlGaInP heterostructures with multiple quantum wells can further generation of impurity compounds (Mg-O, Te-O) both in active and in barrier regions under ionizing irradiation, as the consequence, centers of non-radiative recombination and/or absorption of generated radiation grow

[16]. The presented facts should be taken into consideration for studying radiation-induced defects in the heterostructures and devices made of them.

Quantum wells, introduced into the core of a heterostructure, facilitate concentration growth of minor charge carriers in it due to the state of two-dimensional electronic gas; it increases internal quantum efficiency [1]. Here, elastic strain relaxation on the layers boundary arising due to the inconsistency of lattice parameters with the introduction of quantum wells in the AlGaInP heterostructures active region doesn't lead to dislocations or point defects formation, which can be centers of non-radiative recombination.

At present, the change in parameters of AlGaInP heterostructures with multiple quantum wells exposed to ionizing irradiation has not been adequately investigated [17, 18]. The issue of radiation stability of the LEDs produced on their base is gaining its urgency because they can be used in conditions of various ionizing irradiation [19, 20].

The purpose of the work is to research watt and volt characteristics changes of LEDs based upon AlGaInP heterostructures with multiple quantum wells under fast neutron irradiation and gamma-irradiation.

2. Experimental

The objects of research were red and yellow LEDs (batch LED1) produced on the base of AlGaInP heterostructures with 10 multiple quantum wells (λ = 624 nm) and yellow LEDs (batch LED2) (λ = 590 nm) made by metalorganic chemical vapour deposition method. Dimethylditelluride (DETe) and bis (cyclopentadienyl) magnesium (Cp, Mg) were used as the doping sources for *n*-type and *p*-type layers, respectively. Therefore, under investigation the LEDs were distinguished by the broad band gap, that is, composition of the active region.

At first, a *p*-layer of the structure was dielectrically SiO_2 coated (0.5 μm) and a part of the substrate was chemically and chemical-dynamically etched to obtain the total thickness of the structure (200-210 μm).

Nextly, an ohmic contact was formed towards n-GaAs substrate all over the surface of plates via magnetron-coated Au-Ge-Ni^+-Au deposition (0.15–0.2 μm). An ohmic contact was burnt into in the atmosphere of hydrogen at temperature (425–430) ^{0}C during 5 min.

To form ohmic contacts to *p*-layers of the structures photolotigraphy was carried out on the plates, SiO_2 was etched and after exposed surface of the surface had been chemically treated layers of metals Au-Zn^+-Au (about 0.2 μm) or Ti-Au (0.15–0.2 μm) were coated subsequently or into spaces in photoresist. After coating had been completed photoresist was explosively removed, followed by burning into metal layers in hydrogen atmosphere at temperature 460 ^{0}C for 5 min. In this way mesastructures – prototypes of crystals (chips) were formed. Then contacts to *n*- and *p*-layers of heterostructures were additionally Au-deposited (4–5 μm) (all over the substrate surface and photolithographically formed mesastructures over the surface of *p*-layer).

The crystals were mounted into corresponding cases via thermal-compression assembly.

Fast neutrons with effect level characterized by particle fluence F_n [n/cm^2] and gamma-rays characterized by dose $D\gamma$ [Gy] were chosen as a source of radiation action.

LEDs were irradiated in standard conditions in passive power mode that is without forward current flowing. For the purpose of research different LED sets were made, and each set included at least 20 diodes. For various LED sets exposure level was augmented via sequentially increased radiation dose.

Watt and volt characteristic was measured before and after irradiation in a sphere. In the all experiments focused on irradiation particle flux was directed perpendicularly to the substrate plane. Each experimental value of the below given dependences conforms to a mean parameter value for LED set.

We applied both one-time and sequential sets of required levels of exposure. By comparing the obtained results we can conclude that methods to measure the change in LED parameters don't cause annealing of introduced radiation defects.

3. Results and considerations

Let us consider the results of measurement, which were obtained before irradiation of LEDs. To begin our consideration we analyze the straight branch of volt-ampere characteristics shown in Figure 1. Forward branches of volt-ampere characteristics are in line with the standard LED characteristic with p-n junction at forward bias. Inaccuracy of measurements didn't exceed 3%. Inaccuracy is depicted in Figure 1 as confidence intervals to assess measurement validity. Therefore, batches LED1 and LED2 under consideration demonstrate typical volt-ampere characteristics, which are quite similar to dependences presented in publications [10].

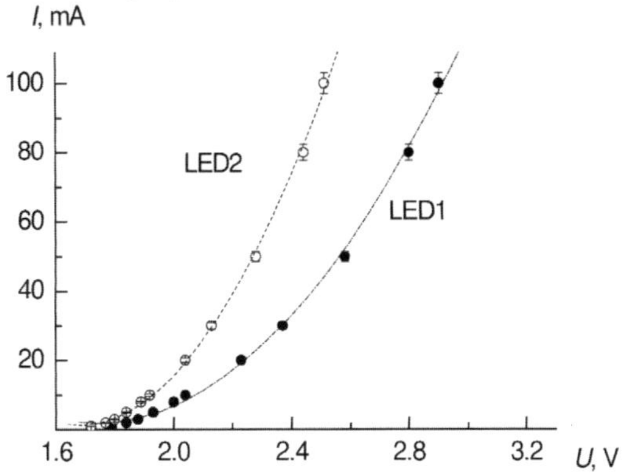

Figure 1. Volt-ampere characteristic of LED1 and LED2 in the linear scale, where symbols are experimental data; lines are identified regularities.

The higher supply voltage is observed for LED2 with identical forward current as compared with LED1 due to the ohmic contacts bad quality which has led to decrease of additional supply voltage in case of LED2.

Watt and volt characteristics are ones of main LEDs characteristics, which determine operating features of an electronic appliance. Relying on the analysis of watt and volt characteristics optimal parameters of appliance maintenance can be selected including voltage, which provides maximum external LED quantum efficiency.

At first, let us consider what the changes in watt and volt characteristics under neutron irradiation are registered. As fast neutron fluence was increased the change in watt and volt characteristics was identified (Figure 2). The light output power is shown in the figures in relative units. Therefore, the change of the results obtained by irradiation you can trace. Neutron flux values reach 1015 n/cm^2.

As one can see, three zones are highlighted for reference diodes, and there is a boundary area between them (selected triangle area patterned by touches in Figure 2). Each pronounced zone has its specific dependence of analyzed regularities.

Therefore, LED external quantum efficiency depends on the different factors as same as the manufacturing technology, chemical composition of material and rate of emitting and full recombination of charge carriers, lifetime of radiative and non-radiative recombination [21–23]. More efficient LED1 (in our case) is distinguished by shorter lifetime of minor charge carriers.

Three typical areas (I, II and area III) can be highlighted for each LED type according to obtained watt and volt characteristics. We can sum up that highlighted areas are those of average electron injection, high electron injection and low electron injection into the LEDs active region is highly pronounced. The frontier between average electron injection and high electron injection is specified by equal concentration of injected electrons and holes in p-region. Our assessment has revealed that concentration of injected electrons is really comparable with concentration of holes in p-region in conditions of the identified frontier.

The dependence of emissive power on supply voltage can be described by power function. Then power exponent in low electron injection area is approached to zero.

IOP Conf. Series: Materials Science and Engineering **363**(2018) 012010 doi:10.1088/1757-899X/363/1/012010

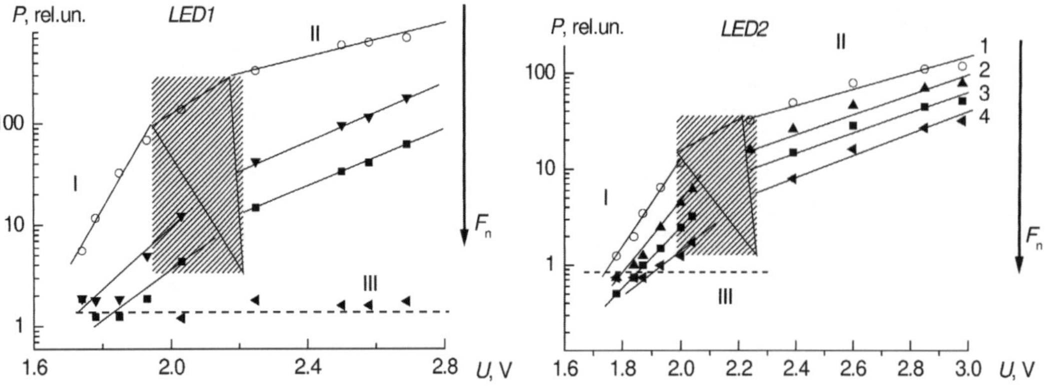

Figure 2. Watt and volt characteristic of the LEDs under irradiation by fast neutrons, where are I is average electron injection area; II is high electron injection area; III is low electron injection area; symbols are results of measurement; lines are identified regularities; *LED1*: 1 – F_n=0; 2 – $F_n=1.07\cdot10^{14}$ n/cm^2; 3 – $F_n=2.74\cdot10^{14}$ n/cm^2; 4 – $F_n=2.57\cdot10^{15}$ n/cm^2; *LED2:* 1 – F_n=0; 2 – $F_n=9.1\cdot10^{12}$ n/cm^2;
3 – $F_n=2.06\cdot10^{13}$ n/cm^2; 4 – $F_n=3.83\cdot10^{13}$ n/cm^2.

At the same time, power exponent in average electron injection zone is always greater than power exponent of high electron injection area. Perhaps this is due to appearance of thermal effects in high electron injection area, whereas noticeable crystal heating by operating current is exactly observed in this zone during taking measurement of watt-ampere characteristics. Temperature increasing of active region leads to quantum efficiency decreasing, which consequence declines power exponent reduction in the high electron injection area.

The transition field is observed between average and high electron injection areas, and it decreases with neutron fluence increasing.

LED1 power exponent is practically irrespective from neutron fluence in high electron injection area. It noticeable decreases in the average electron injection area. In the limiting case under neutron irradiation low electron injection area initially extends into the average electron injection area, and then on the high electron injection area.

Let us consider the change in watt-voltage characteristics of AlGaInP LEDs under gamma-irradiation (Figure 3).

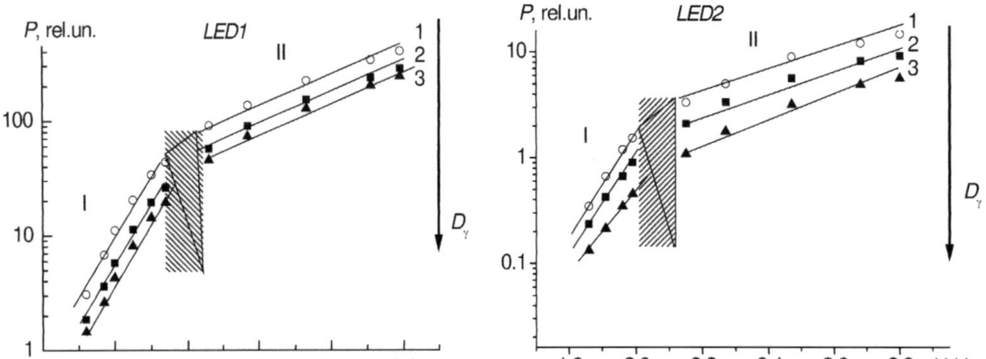

Figure 3. Watt and volt characteristic of the LEDs under gamma-irradiation, where I is average electron injection area; II is high electron injection area; III is low electron injection area; symbols are results of measurement; lines are identified regularities; *LED1*: 1 – D_γ=0; 2 – $D_\gamma=4.18\cdot10^7$ Gy; 3 – $D_\gamma=1.08\cdot10^8$ Gy; *LED2*: 1 – D_γ=0; 2 – $D_\gamma=1.76\cdot10^8$ Gy; 3 – $D_\gamma=4.68\cdot10^8$ Gy.

The change of the obtained results under irradiation by ^{60}Co gamma-rays shows a decrease in the relative power of the radiation due to increasing exposure. The gamma irradiation dose values reach

10^8 Gy. These results show that observed changes of watt and volt characteristic under gamma-irradiation are identity which is observed previously under neutron irradiation (Figure 2).

Previously [17–20], we have established that LEDs light output power at predetermined operating current has decreased as a result of ionizing radiation in three stages. On the first stage emissive power reduction is the result of radiation-enhanced existing defectiveness restructuring. Present stage characterizes by light output power decrease saturation with exposure level increase.

On the second stage emissive power decreases by exclusively radiation defects. On the third stage LEDs goes into low electron injection area. At the same time emissive power is practically independent from the operating current value with its changes in investigated current ranges. The third stage characterizes ultimate resistance of observable LEDs.

The first stage contribution of the emissive power decreasing under irradiation can be characterized at percentage ratio by value A and the second stage contribution – value B, so (A + B) = 100%.
In the analysis of available experimental data of LED1 and LED2 we have determined following ratio for fast neutrons:

$$A_n = \left(\frac{I_b}{I_{op}}\right)^{\alpha_n} \times 100\% \text{ ,} \quad (1)$$

where A_n is first stage contribution of the emissive power decreasing under neutron irradiation, %; I_b is operating current value measured on the boundary between low and average electron injection areas; I_{op} is operating current value, which is considered the changes of emissive power under irradiation; $\alpha_n = 0.28$ is power exponent for fast neutrons.

For gamma radiation we have established similar ratio:

$$A_\gamma = \left(\frac{I_b}{I_{op}}\right)^{\alpha_\gamma} \times 42\% \text{ ,} \quad (2)$$

where A_γ is first stage contribution of the emissive power decreasing under gamma irradiation, %; $\alpha_\gamma = 0.11$.

To compare ratio (1) with ratio (2) we can make the conclusion that gamma-irradiation can be able to reconstruct only part (42 %) of initial defect structure in comparison with fast neutrons.

Analysis of established regularities allows to state that LEDs operating current level on the boundary between low and average electron injection areas is determined by used semiconductor material defectiveness. Also it defines resistance to fast neutrons and ^{60}Co gamma-rays irradiation.
So the correlations (1, 2) allow us to predict radiation resistance of investigated LEDs.

4. Conclusion

In conclusion main results of investigation into deterioration of watt and volt characteristics of AlGaInP LEDs based on heterostructures (λ1 = 590 nm and λ2 = 630 nm) with multiple quantum wells under fast neutron and gamma irradiation are summarized:

1. Watt-voltage characteristics of LEDs based on AlGaInP heterostructures with quantum wells in the average and high electron injection areas can be described as a power function of the operating voltage. Wherein the power exponent in average electron injection is always higher than in a high electron injection area. In the low electron injection area exponent is zero, i.e. LEDs emission power does not depend on the supply voltage.

2. Transition field is observed on the boundary between the average and high electron injection areas. It disappears as you get closer to the limit result of the irradiation LEDs - low electron injection mode in the entire supply voltage range.

3. During irradiation with the fast neutron fluence growth (with increasing gamma rays absorbed dose) exponent in the average electron injection area is reduced. It is equal to zero in the limit. While high electron injection area extends into the average electron injection area.

4. At the beginning irradiation with the fast neutron fluence growth (with gamma rays absorbed dose increasing) power exponent in the high electron injection area is practically independent of exposure level (area low impacts – the existence region of the average electron injection mode). And then it decreases and equals zero in the limit, while the low injection area extends into the high electron injection.

5. Measured on the boundary between low and average electron injection areas current value depends on the initial defectiveness of the semiconductor material. It determines first stage contribution of LEDs light output power reducing under irradiation due to radiation-enhanced initial defects restructuring in total emissive power decreasing.

6. The first stage contribution of emissive power reducing under irradiation by gamma rays lower (42%) than the observed fast neutrons contribution (100%).

7. Established ratio between measured on the boundary between low and average electron injection areas current value and the contribution magnitude of the first stage LEDs emissive power reducing allows to predict their resistance to irradiation by fast neutrons and gamma rays.

References

[1] Hodapp M W High brightness light emitting diodes 1997 *New York: NY, Academic press* 87–92 doi: 10.1016/S0080-8784(08)62407-2

[2] Plotnikova I V *et al* 2018 *IOP Conf Ser: Mater Sci Eng* **289** **(1)** 012029 doi: 10.1088/1757-899X/289/1/012029

[3] Plotnikova I *et al* 2018 *MATEC Web of Conf* **155** 01052 doi: 10.1051/matecconf/201815501052

[4] Galtseva O V *et al* 2016 *IOP Conf Ser: Mater Sci Eng* **110** 012094 doi: 10.1088/1757-899X/110/1/012094

[5] Gavrilin A *et al* 2016 *MATEC Web of Conf* **79** 01078 doi: 10.1051/matecconf/20167901078

[6] Craford G M 2000 *MRS bulletin* **25(10)** 27-31 doi: 10.1557/mrs2000.200

[7] Corbett B *et al* 2003 *Proceedings of SPIE* **4876** 176-183 doi: 10.1117/12.463713

[8] Zeller H R 1995 *Solid-State Electronics* **38(12)** 2041-2046 doi: 10.1016/0038-1101(95)00082-5

[9] Sugawara H *et al* 1992 *Applied physics letters* **61(15)** 1775-1777 doi: 10.1063/1.108423

[10] Feng Z C *et al* 1999 *Journal of Applied Physics* **85(7)** 3824-3831 doi: 10.1063/1.369752

[11] Bemš J et al 2014 Radiation Physics and Chemistry **104** 398–403 doi: 10.1016/j.radphyschem.2014.02.008

[12] Kapranov B I *and* Varga V V 1999 *Proc of KORUS'1999* **2** 876255 672–676 doi: 10.1109/KORUS.1999.876255

[13] Alkhimov Yu V *et al* 2012 *Russian Journal of Nondestructive Testing* **48(4)** 238–244 doi: 10.1134/S106183091204002X

[14] Vavilova G *and* Ryumkin A 2018 *IOP Conf Ser: Mater Sci Eng* **289(1)** 012017 doi: 10.1088/1757-899X/289/1/012017

[15] Surzhikov A P *et al* 2016 *IOP Conf Ser: Mater Sci Eng* **110(1)** 012002 doi: 10.1088/1757-899X/110/1/012002

[16] Maranowski S A *et al* 2005 *Proc of SPIE* **3002/110** 110–118

[17] Gradoboev A V *and* Orlova K N 2015 *Physica Status Solidi C* **12(1-2)** 35–38 DOI: 10.1002/pssc.201400072

[18] Gradoboev A V *and* Orlova K N 2014 *Advanced Materials Research* **880** 237–241 DOI: 10.4028/www.scientific.net/AMR.880.237

[19] Orlova K N *et al* 2012 *Proc of IFOST'2012* **1** 1-4 doi: 10.1109/IFOST.2012.6357528

[20] Orlova K N *and* Gradoboev A V 2014 *Proc of CriMiCo'2014* 874-875 doi: 10.1109/CRMICO.2014.6959672

[21] Ashry M *et al* 2004 *Radiation Effects and Defects in Solids* **159(8-9)** 453-460 doi: 10.1080/10420150410001670297

[22] Suzuki M *et al* 1993 *Journal of Crystal Growth* **133**(3-4) 303-308 doi: 10.1016/0022-0248(93)90169-W

[23] Makhkamov S *et al* 2005 *Radiation Effects and Defects in Solids* **160(8)** 349–356 doi: 10.1080/10420150500415116

System model the processing of heterogeneous sensory information in robotized complex

V Nikolaev[1], V Titov[1], V Syryamkin[2]
[1] South-West state University
[2] National Research Tomsk State University, 634050, Tomsk, Russia
E-mail: nikovic54@yandex.ru

Abstract. Analyzed the scope and the types of robotic systems consisting of subsystems of the form "a heterogeneous sensors data processing subsystem". On the basis of the Queuing theory model is developed taking into account the unevenness of the intensity of information flow from the sensors to the subsystem of information processing. Analytical solution to assess the relationship of subsystem performance and uneven flows. The research of the obtained solution in the range of parameter values of practical interest.

1 Introduction.

The modern stage of society development is characterized by the expansion of application fields of robototechnic complexes (RC) and technology. The basis for the structure of RK is a system of diverse sensors, performing the measurement, collection, formation and direction of the subsystem of information processing of different parametric form [1,2].

The most common in Kazakhstan are internal local area network (LAN) of the form "a heterogeneous sensors data processing subsystem". Such LVS are always an integral part of automated systems and Kazakhstan in different areas of arts activities: research; education; industry; economy; medicine; the social sphere [3–8].

In this formulation, important and urgent is to develop a mathematical model that allows to evaluate and formulate requirements to the parameters of the model elements of such a private LAN.

Theoretical part. The analysis of literary sources shows that in about-setitemnum plan, the basic elements of the internal LAN RK consist of the following main elements [2]:

1. Heterogeneous sources – sensors (sensors) i.e., the initial elements of the radiation data or information about the state or properties of a process, technology or RK.

2. Receiver (subsystem of information processing) of the initial elements included in the means of measurement and monitoring information and the feature space.

The processes of functioning of systems with such structural and functional organization most adequately describes the models and methods of Queuing theory [9].

Therefore, for solving the problem develop a model of the functioning of the internal LAN of Kazakhstan it is necessary to investigate the queueing system of type MP/M/1//N, where N elements of the lower level in the form of working sensors initiate service requests with different intensity, i.e., as it is over-hanging on the type of sensor and solved RK at the moment of the task.

Graphs of the functioning of the LVS RK a closed type with multiple sensors having different intensity of information flow and a processing engine is shown in Figure 1.

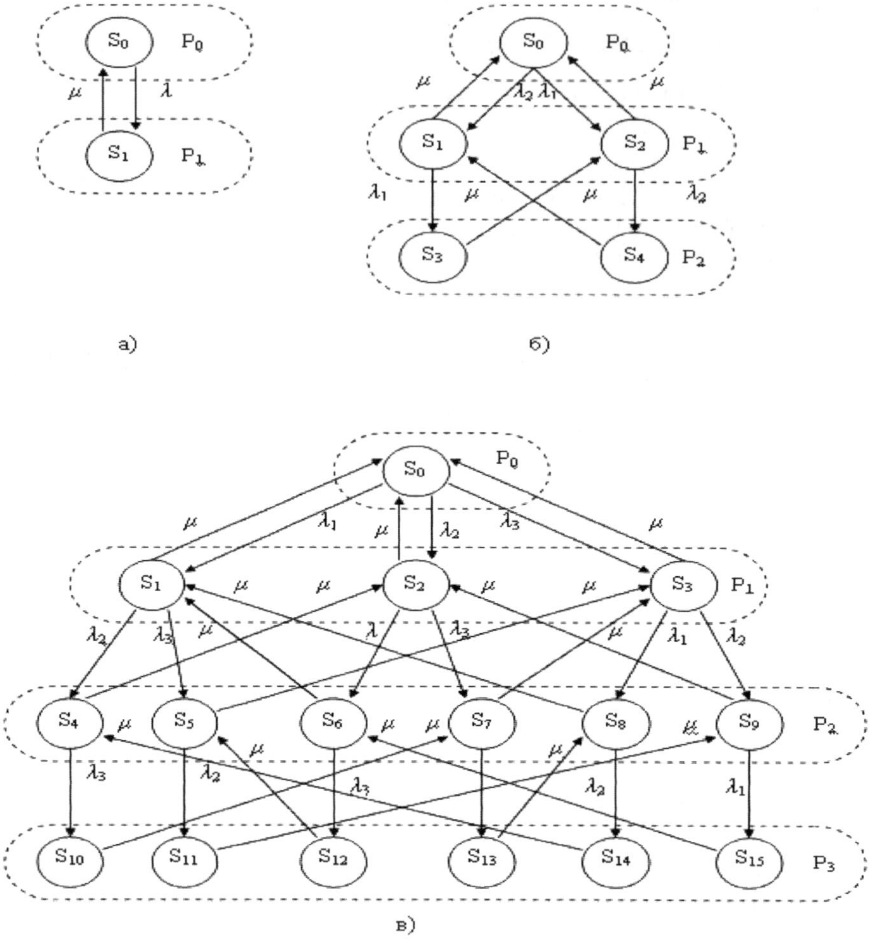

Figure 1. Graphs of the functioning of the LVS RK a closed-type sensors having different intensity of information flows: a) N=1; b) N=2; a) N=3.

Note that the solution for the model of type M/M/1//N, according to the classification of Kendall, ie, with the same intensity of service is known and the difference of the proposed model consists of recording the fact of non-uniformity of generation of the information flow sensors of the lower level.

In the first stage, first decide the task for the special cases N=1, N=2, N=3 and try to identify patterns, which can be used to find a common solution.

Based on the graphs depicted in Figure 1 and describing the process of functional planning of such a system, for different occasions make a system of linear differential equations [10] that will move on to the differential. These equations have the following form.

$N = 1$

$$\begin{cases} \mu S_1 - \mu S_0 = 0; \\ \lambda_1 S_0 - \mu S_1 = 0; \\ \sum_{i=0}^{1} S_i = 1. \end{cases}$$

$N = 2$

$$\begin{cases} \mu S_1 + \mu S_2 - (\lambda_1 + \lambda_2) S_0 = 0; \\ \mu S_4 + \lambda_2 S_0 - (\lambda_1 + \mu) S_1 = 0; \\ \mu S_3 + \lambda_1 S_0 - (\lambda_2 + \mu) S_2 = 0; \\ \lambda_1 S_1 + \mu S_3 = 0; \\ \lambda_2 S_2 + \mu S_4 = 0; \\ \sum_{i=0}^{4} S_i = 0. \end{cases}$$

$N = 3$

$$\begin{cases} \mu S_1 + \mu S_2 + \mu S_3 - (\lambda_1 + \lambda_2 + \lambda_3) S_0 = 0; \\ \mu S_6 + \mu S_8 + \lambda_1 S_0 - (\lambda_2 + \lambda_3 + \mu) S_1 = 0; \\ \mu S_4 + \mu S_9 + \lambda_2 S_0 - (\lambda_1 + \lambda_3 + \mu) S_2 = 0; \\ \mu S_5 + \mu S_7 + \lambda_3 S_0 - (\lambda_1 + \lambda_2 + \mu) S_3 = 0; \\ \mu S_{14} + \lambda_2 S_1 - (\lambda_3 + \mu) S_4 = 0; \\ \mu S_{12} + \lambda_3 S_1 - (\lambda_2 + \mu) S_5 = 0; \\ \mu S_{15} + \lambda_1 S_2 - (\lambda_3 + \mu) S_6 = 0; \\ \mu S_{10} + \lambda_3 S_2 - (\lambda_1 + \mu) S_7 = 0; \\ \mu S_{13} + \lambda_1 S_3 - (\lambda_2 + \mu) S_8 = 0; \\ \mu S_{11} + \lambda_2 S_3 - (\lambda + \mu) S_9 = 0; \\ \lambda_3 S_4 - \mu S_{10} = 0; \\ \lambda_2 S_5 - \mu S_{11} = 0; \\ \lambda_3 S_6 - \mu S_{12} = 0; \\ \lambda_1 S_7 - \mu S_{13} = 0; \\ \lambda_2 S_8 - \mu S_{14} = 0; \\ \lambda_1 S_4 - \mu S_{15} = 0; \\ \sum_{i=0}^{14} S_i = 1. \end{cases}$$

For the stationary regime we get the following solutions:

$N = 1$:

$$P_0 = S_0 = \frac{1}{\rho_1 + 1}; \qquad (1)$$

$$P_1 = S_1 = \frac{\rho_1}{\rho_1 + 1}.$$

$N = 2$:

$$P_0 = S_0 = \frac{1}{(\rho_1 \rho_2 + \rho_2 \rho_1) + (\rho_1 + \rho_2) + 1};$$

$$P_1 = S_1 + S_2 = \frac{\rho_1 + \rho_2}{(\rho_1\rho_2 + \rho_2\rho_1) + (\rho_1 + \rho_2) + 1};$$

$$P_2 = S_3 + S_4 = \frac{\rho_1\rho_2 + \rho_2\rho_1}{(\rho_1\rho_2 + \rho_2\rho_1) + (\rho_1 + \rho_2) + 1}.$$

$N = 3$:

$P_0 = S_0 = 1/ [(\rho_1 \rho_2 \rho_3 + \rho_1 \rho_3 \rho_2 + \rho_2 \rho_1 \rho_3 + \rho_2 \rho_3 \rho_1 + \rho_3 \rho_1 \rho_2 + \rho_3 \rho_2 \rho_1 + \rho_1 \rho_2 + \rho_1 \rho_3 + \rho_2 \rho_3 + \rho_2 \rho_1 + \rho_3 \rho_1 + \rho_3 \rho_2) + (\rho_1 + \rho_2 + \rho_3) + 1]$;

$$P_1 = \sum_{i=1}^{3} S_i = (\rho_1 + \rho_2 + \rho_3)/[(\rho_1\rho_2\rho_3 + \rho_1\rho_3\rho_2 + \rho_2\rho_1\rho_3 + \rho_2\rho_3\rho_1 + \\ + \rho_3\rho_1\rho_2 + \rho_3\rho_2\rho_1) + (\rho_1\rho_2 + \rho_1\rho_3 + \rho_2\rho_3 + \rho_2\rho_1 + \rho_3\rho_1 + \rho_3\rho_2) + \\ + (\rho_1 + \rho_2 + \rho_3) + 1];$$

$$P_2 = \sum_{i=4}^{9} S_i = (\rho_1\rho_2 + \rho_1\rho_3 + \rho_2\rho_3 + \rho_2\rho_1 + \rho_3\rho_2)/[(\rho_1\rho_2\rho_3 + \rho_1\rho_3\rho_2 + \\ + \rho_2\rho_1\rho_3 + \rho_2\rho_3\rho_1 + \rho_3\rho_1\rho_2 + \rho_3\rho_2\rho_1) + (\rho_1\rho_2 + \rho_1\rho_3 + \rho_2\rho_3 + \rho_2\rho_1 + \\ \rho_3\rho_1 + \rho_3\rho_2) + (\rho_1 + \rho_2 + \rho_3) + 1];$$

$$P_3 = \sum_{i=10}^{15} S_i = (\rho_1\rho_2\rho_3 + \rho_1\rho_3\rho_2 + \rho_2\rho_1\rho_3 + \rho_2\rho_3\rho_1 + \rho_3\rho_1\rho_2 + \\ + \rho_3\rho_2\rho_1)/[(\rho_1\rho_2\rho_3 + \rho_1\rho_3\rho_2 + \rho_2\rho_1\rho_3 + \rho_2\rho_3\rho_1 + \rho_3\rho_1\rho_2 + \rho_3\rho_2\rho_1) + \\ + (\rho_1\rho_2 + \rho_1\rho_3 + \rho_2\rho_3 + \rho_2\rho_1 + \rho_3\rho_1 + \rho_3\rho_2) + (\rho_1 + \rho_2 + \rho_3) + 1];$$

where Si is a discrete state Markov chain describing the functioning of the regulation of LVS RK;

RK group of States, which determines the probability of finding a system To the requests;

$\rho_i = \lambda_i / \mu$ – utilization of the system i-stream applications;

$\lambda_i = 1 / \bar{t}^{(1)}_{cpi}$ – the intensity of the flow of applications from the i-sensor RK;

$\mu = 1 / \bar{t}^{(2)}_{cp}$ – intensity of service requests in the engine.

The solution of the problem for cases N > 3 becomes much more complicated, however, analysis of the structural expression (1) allows to obtain a General formula for determining RK for any number of sensors having different intensity of the appearance of the queries [11]:

$$P_0 = \left[\sum_{K=0}^{N} A_N^K \prod_{i \in N} \rho_i^{(M)} \right]^{-1}; \tag{2}$$

$$P_K = \begin{cases} P_0 A_N^K \prod_{i \in N} \rho_i^{(M)}, & 0 < K < N, \\ 0, & K > N, \end{cases} \tag{3}$$

where $M = A_N^K$ – he number of placements of N by K;

$\prod_{i \in N} \rho_i^{(M)}$ – the work of a group of parameters PI that are included in the M-accommodation.

The equation (2,3) is obtained by deduction based on the detection of naturally-STI in changing the coefficients in the equations and their extrapolation to the General case. Additionally, the accuracy and correctness of the obtained solution is checked-Las by analyzing the condition $\sum_{K=0}^{N} P_k = 1$ on the grid of parameters whose values changed in the range of practical interest for this class of subsystems of the processing in RC. In this case, N = 1(1)100 and PN = 10-3(10)103. Inspection of the expressions (2-3), are made nye by means of simulation, showed their accuracy and their applicability for the analysis of queueing systems of the closed type with various elements.

To assess the impact of the uneven information processing sensors the lower level to the time parameters of the processing subsystems will conduct a research of the solution. Thus comparable values for the utilization rate of system resources subsystem processing options with identical and different treatment times, but with equal total intensity of the stream of queries processed, i.e.,

$\lambda_{об} = N\lambda$ – for RK with the same sensors;

$\lambda_{об} = \sum_{i=1}^{N} \lambda_i$ – for RK with different sensor sources.

Analyze in the framework of the approach [12, 13] will change as the parameter $\overline{L}_a (\overline{L}_a = 1 - P_0)$, defining bandwidth $B = \mu \overline{L}_a$, in the framework of the proposed model. First of all, note that the same number of sensors and is equal to the total intensity $\overline{L}_a$ will be different as it is possible to obtain a large number of values works $\prod_{i \in N} \rho_i^{(M)}$ when changing the parameters PI. So important is the determination of possible boundaries changes.

The maximum value $\overline{L}_a$ accept in case of equality of intensity of all sources:

$$\overline{L}_{a_{\max}} = 1 - \left[\sum_{K=0}^{N} \rho^K \frac{N!}{(N-K)!} \right]^{-1}. \tag{4}$$

Minimal, if $\rho_N \to \rho_{об}$, a $\sum_{i=1}^{N} \rho_i \to 0$, i.e. when the intensity of one source is much higher than the total intensiveness of the rest. Then the expression for $\overline{L}_a$ has the form:

$$\overline{L}_{a_{\min}} = 1 - \left[\sum_{i=1}^{N} \rho_i + 1 \right]^{-1}. \tag{5}$$

Consider the amount by which it is possible to change $\overline{L}_a$. For this we introduce the function $\delta(N, \rho_{об}) = \overline{L}_{a_{\max}} - \overline{L}_{a_{\min}}$, reflecting the absolute deviation of the values $\overline{L}_a$ in the case of N – sources with the same intensity and in the worst case with N – sources of varying intensity:

$$\delta(N, \rho_{об}) = \left[\sum_{i=1}^{N} \rho_i + 1 \right]^{-1} - \left[\sum_{K=0}^{N} \rho^K \frac{N!}{(N-K)!} \right]^{-1} \tag{6}$$

The value of the function $\delta(N, \rho_{об})$ for different N is shown in Figure 2.

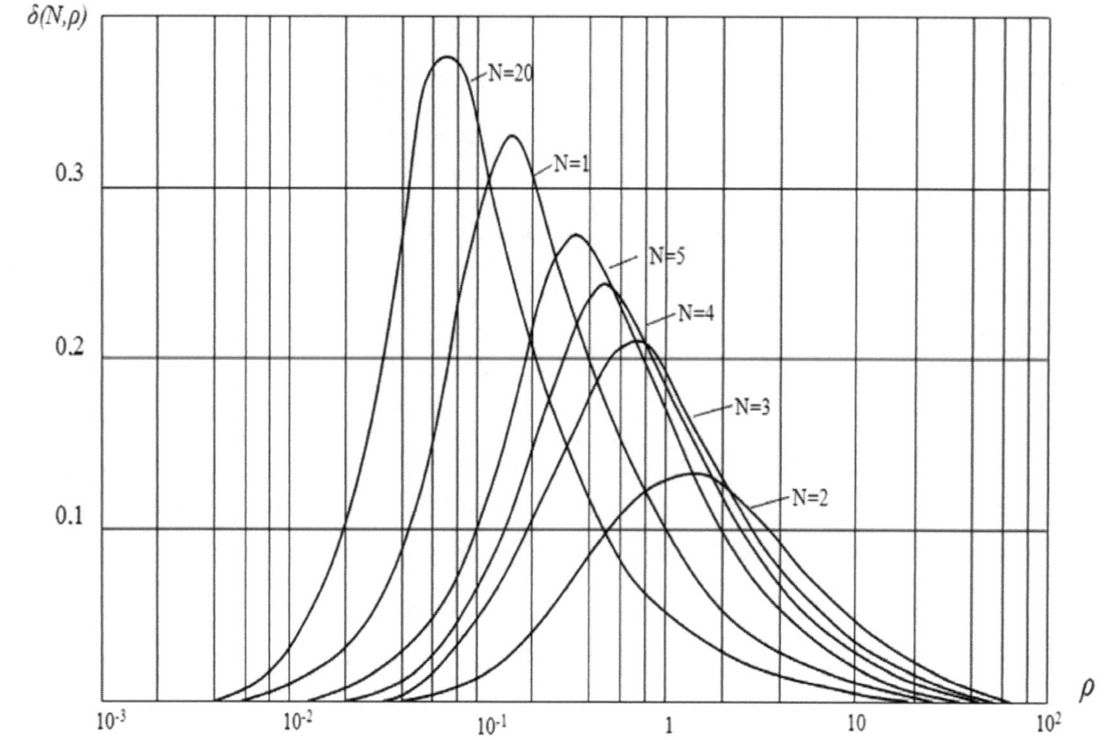

Figure 2. Schedules of values δ(N, ρ)

As $0 \le \overline{L}_a \le 1$, it $\delta(N, \rho_{об})$ a good illustration of the change $\overline{L}_a$. For different N it is possible to accurately find the maximum value of δ(N, Rob) by determining the first derivative of expression (6) and then solving the resulting equation for ρ. However, because of the cumbersome decision appropriate to use numerical methods.

The analysis of graphs in Fig. 2 shows that the greatest discrepancies are observed $6{\cdot}10^{-3} \le \rho_{об} \le 20$, where δ $(N, \rho_{об})$ varies from 0.05 to 0.35.

In $\rho_{об} > 20$ and $\rho_{об} \le 6{\cdot}10^{-3}$ the differences are minor and do not exceed 12%. The observed increase in the discrepancy function δ(N Rob) as the number of elements, while at the same time the maximum moves to the left.

2 Conclusion

The studies show that the value of the parameter $\overline{L}_a$ in the two-phase model of mass service closed-circuit, originating in matter from sensors with different intensity, with different parameters N, ρ, may be different from the known models by more than 35 %. Given the fact that the parameter $\overline{L}_a$ directly determines the bandwidth of this structure, it is obvious that for the analysis of processes of functioning of the internal LAN of Kazakhstan it is necessary to use the developed model.

Acknowledgments

This study (research grant No № 8.2.24.2018) was supported by the Tomsk State University competitiveness improvement programme.

IOP Conf. Series: Materials Science and Engineering **363** (2018) 012011 doi:10.1088/1757-899X/363/1/012011

References

[1] Myasnyankina D V, Nikolaev V N 2014 *Information-measuring and control system* **(8)** 46–49
[2] Nikolaev V N, Stucco A V 2016 *MATEC Web of Conferences* 01042
[3] Nikolaev V N, Titov G S 2013 *Medical devices* **(1)** 245–249
[4] Nikolaev V N 2012 *Instrumentation* **55**(2) 55–58
[5] Maslak A A, Nikolaev V N 2013 *Information-measuring and control system* **11**(8) 46–49
[6] Nikolaev V N, Dorokhov D S, Tovin A E 2015 *Innovation economy: prospects for development and improvement* **2 (7)** 192–195
[7] Nikolaev V N *et al* 2013 *Telecommunications and Radio Engineering* **72**(13)1219–1225
[8] Korenevsky N A *et al* 2016 *News of southwest state University. Series: Management, computer engineering, computer science. Medical devices* **2 (19)** 136–142
[9] Kalashnikov V, Rachev S T 1988 *Mathematical methods for constructing STO-khastichenka systems service models* (Moscow, Nauka)
[10] Buslenko N P1968 *Modeling of complex systems* (Moscow, Nauka)
[11] Nikolaev V N 2005 *Control Systems and information technology* **(3)**. 70–74
[12] Nikolaev V N 2006 *Control Systems and information technology* **(3)** 19–22
[13] Yurchenko A *et al* 2016 *MATEC Web Conf.* **79** 01085

Cognitive and sociocultural aspects of robotized technology: innovative processes of adaptation

S B Kvesko[1], B B Kvesko[2], M A Kornienko[1], Y A Nikitina[3] and N M Pankova[3]
[1]Research Tomsk State University, Tomsk, Russia
[2]Siberian Federal University, Krasnoyarsk, Russia
[3]National Research Tomsk Polytechnic University, Tomsk, Russia
E-mail: svetla_kvesko@mail.ru

Absract. The paper dwells upon interaction between socio-cultural phenomena and cognitive characteristics of robotized technology. The interdisciplinary approach was employed in order to cast light on the manifold and multilevel identity of scientific advance in terms of robotized technology within the mental realm. Analyzing robotized technology from the viewpoint of its significance for the modern society is one of the upcoming trends in the contemporary scientific realm. The robots under production are capable of interacting with people; this results in a growing necessity for the studies on social status of robotized technological items. Socio-cultural aspect of cognitive robotized technology is reflected in the fact that the nature becomes 'aware' of itself via human brain, a human being tends to strives for perfection in their intellectual and moral dimensions.

1. Introduction

Within this framework, the viewpoint analysis is argued to be significant, as well as the multidisciplinary approach to the issue of robotized technology and the general outline for the development prospects of the latter.
Relying on the determinist perspective, the present paper strives:
1) to define the theoretical and methodological background of robotized technology phenomenon;
2) to state the chief developmental paths for robotized technology within the social environment taking into account robot vs. human interactive processes and the application sphere.

The 21st century scientific and technologic standards stipulate the existence of a certain communicational space that allows to adjust robotized technology to a human's abilities [1]. In spite of the fact that robotized technology system involves mainly automated digital items, it is still able to compensate the deficiency of natural human capabilities, which bears axiological significance of manmade civilization.
Cognitive robotized technology allows mitigating man-power deficiency. Artificial intelligence plays a significant role in the formation and development of robotized technology realm, the former generating a new status of the latter via synthesizing robotized technology with cognitive robotized technology.

The aim of the present research is to analyze the socio-cultural aspect of the cognitive robotized technology. The aforementioned aim implies the formulation of numerous objectives, namely:
1) analyzing the essence of cognitive robotized technology as a system of robot vs. human interaction within a social environment for purposes;
2) exploring the socio-cultural aspect of robot vs. human interaction.

Aims and objectives were reached by means of socio-cultural method and cognitive approach. Socio-cultural method implies analyzing society as a single entity of culture and humankind social relations, while a single individual is understood to be a micro-system of culture and relationship, as well as an amalgamation of social values and norms. In turn, cognitive approach implies taking into consideration the importance of knowledge for better understanding of human behavioral patterns. Besides, synergetic and analytical methods have been employed, as well as the competence-based approach to the analysis of robotized technology as one of the most important issues in the sphere of social engineering.

2. Cognitive and sociocultural factors of robotized technology as the innovation

From the standpoint of the cognitive approach, robot presents an epistemological model. Being a product of the social construction process, epistemological robotic models are to a certain extent the replicas of human consciousness and activity. Robotized technology in cognitive realm integrates the epistemological characteristics into socio-cultural values.

The coordination of epistemological and socio-cultural factors provides the society with multidimensionality of cognitive robotized technology, that is, while analyzing the essence of cognitive robotized technology, we observe a convincing contrast between descriptive and analytical methods and research tools [2].

Modern cognitive robotic models allow us to figure out the mechanisms of their production following worldview systems and scientific traditions, which stimulates the revelation of the new values [3].

The worldview level of technical knowledge predetermines the cognitive context of robotized technology production processes. It is the worldview ideas that guide the processes of research, outlining, and construction within the sphere of artificial intelligence and cognitive robotized technology.

The emerging robotic science possesses the whole framework of diverse conceptual, cognitive, and sociocultural systems, which contributed to the identification of prerequisites for the establishment of cognitive robotized technology. The mentioned prerequisites have been integrated into a cohesive systemic unity of the numerous axiological and cognitive factors.

Cognitive and socio-cultural factors of robotized technology advance predetermine the background and dimensionality for the ways artificial intelligence works. It is impossible to ignore the importance of sociocultural factors (including worldview systems) since they have managed to remove the impediments on the way of non-standard scientific novelties [4].

The sociocultural factors had their influence on cognitive factors (including IT, math, engineering, etc.) and served as the basis for the genesis of new knowledge. Human beings consider themselves to be the highest value of culture; such position bears the name of anthropocentric. This fact heightens man's interest in the production of thinking self-organizing machines – as humankind assumes that they create gnostic technologic methods capable of transforming the world in full compliance with man's desire for environmental changes [5].

Robotized technology is characterized by being oriented towards conservation and sustainable development of intra- and extra-scientific factors that increase the ideological importance of the role the robotized technology plays in scientific advance, educational and manufacturing processes, as well as in the growing social awareness.

Throughout the increasing development of cognitive and humanities, the revision of interdisciplinary approaches to progressive scientific changes takes place. It allows researchers to reconstruct the changes taking into account interdependent cognitive and socio-cultural aspects. In the course of ongoing research, it has been attempted to analyze the multidimensional identity of the interaction between socio-cultural and cognitive factors within the realm of robotized technology advance [6].

Technologic advance provides an opportunity to produce robots capable of solving problems beyond human capacity, which results in the appearance of new and profound prospects for human activity development and improvement.

However, in order to realize these prospects it is required to re-consider the personal factor within society and social foundations.

Performance management for the development of robotized technology (especially in its cognitive branch) should be aimed at innovations in general and getting the staff prepared for the introduction of innovative technologies in particular.

Considering the prospects for the robotized technology development, it should be noted that there exist high risks in the sphere of its implementation from socio-cultural viewpoint. To benefit from the introduction of expensive high-tech robotized technology items and to commensurate with the costs of their production, it is necessary to analyze the indicators, which will stipulate the maximum profit.

While considering the prospects of high-tech robotic technology, it is necessary to carry out analysis of current social and scientific trends. The ensuing outcome will foster further anticipation and predict of social relations transformation.

Sociocultural aspect in cognitive robotized technology reflects the framework of a particular culture. Socio-cultural content owes its significance to scientific, educational, and production tasks, as well as to developmental problems. Such an event results in certain psychic transformation; the development of awareness- and knowledge-based personal qualities begins to take place.

Sociocultural content robotics contributes to the formation of human culture through the formation of social and cultural awareness, as well as by means of conscious approach to robotics. This is a major scientific and educational target in conjunction with developing a person's readiness to soak up the technologic innovations. Therefore, the socio-cultural component becomes a prioritized field of tech advance [7]. The socio cultural factors identity depends on clear understanding of modern society problems.

Social and cultural factor – in order to activate the ideological triggers of the man vs. robot interaction – strives for the familiarization of robotics, the formation of the up-to-date' mentality within the field of science and technology, the promotion of cultural, technical, and socio-cultural enlightenment.

Socio-cultural approach provides information about society, cultural phenomena in various stages of social development. Still, there exists a downside to the impact of socio-cultural factors, which is expressed in a distorted ideological standpoint of society: the social image remains unrealistic; it is to a certain extent idealized – being a result of abstraction and idealization.

During the image formation process, the phenomenon of emerges–resulting in a recurring stereotype, which is becoming increasingly typical for a robotic model. This model only partially reflects the social culture being in a way static, for it includes a set of current and prior knowledge, art, morality, law, etc. [8].

Humankind employs a cognitive model in order to acquire and develop the basic and supplementary expertise. The adjustment to the social culture, to the social reality, and to the social environment serves as the basis for different approaches to socio-cultural component of technical education(including robotized technology).

The ethnographic orientation of socio-cultural approach is to be expressed by means of describing social culture and various ethnic groups demand for modern robotics, as well as the nature of these requirements implementation [9]. The social agent serves as a simulation model, a sort of ideal for creating robotic models capable of meeting the needs of this or that nationality. The purpose of education in robotics is to adapt human beings to the real-life socio-cultural norms of human society.

The pragmatic aspect of sociocultural method concerning the production of robotic devices must be oriented towards understanding of contemporary developmental processes in the field of manufacturing, science, education, which requires the development of completely automated robotics– due to the ongoing formation of IT-domineering society and digital reality [10]. The modern conception of social / cultural impact on the nature of robotics under production implies dialogue-oriented approach based on national, climatic, and other needs of the people [11]. This trend of socio-cultural methodology bears the name of analytical method based on comparative techniques with a special focus on the cultures and culture vs. robotics juxtaposing.

However, civilization is multicultural [12]. Thus, technology – including robotics – can be regarded as a system of multicultural and universal values. For a deeper understanding of cognitive robotized technology, it is necessary to study culture and diverse tech systems. It is strongly recommended that cultural studies, communication theory, and mental pictures become the subject-matter of the research.

Comparative analysis of socio-cultural approach states multiculturalism as its chief postulate. In this respect, great significance role might be attributed to robotic models produced by means of socio-cultural method, since it is cognitive robotics that is capable of transferring information to a single individual and of disseminating it to the public through the prism of cultural image of an event and shaped other-culture' images.

Sociocultural aspect of cognitive robotics lies in the fact that with the help of a robotic model the new cultural reality emerges. As a result, the processes of socialization and cultural self-determination are triggered by means of cognitive robotized technology. Understanding cultural identity and understanding of cultural unity are outstandingly important. People begin to identify themselves as members of society, as they become integrated into technical systems being the results of human intellectual activities(i.e., cognitive robotics, thinking machines, artificial intelligence in general). While choosing an aspect for socio-cultural robotic model it is important to consider interrelated and interacting factors of different cultures, as well as the integration of cultures into the global sociocultural space. Sociocultural content is the basis of communication practice where one cognitive robotics acts as one of the agents [13].

Cognitive robotics – serving as the subject of the communication space and practice – allows us to compare and interpret our attitude to cultural values. It is necessary to classify the sociocultural principles of cognitive robotics. The authors distinguish the following principles [14]: accessibility; universality; efficiency.

Cognitive robotics systems must be available to users, the latter being the stakeholders of communicative space. Socio-cultural aspect of the robot vs. human vs. society communication content may become available in case of robotized systems control language becoming a universal language of communication – clear, discernible and understandable for any subject within social and cultural space [8].

The principle of cognitive robotics efficiency implies the presence of robotic-triggered essential achievements within the spheres of science, education, and manufacturing. Besides, the mentioned achievements must significantly exceed the previous results and allow the world to increase the rating of this particular subject / enterprise.

Sociocultural aspect of cognitive robotic systems presupposes the presence of two forms of human interaction within the space communication socio-cultural space [9], namely [15]:

1) explicit socio-cultural content. It manifests itself through legible / audible texts that dwell upon the socio-cultural events, phenomena, or actions.

2) implicit socio-cultural content that is supplied in the form of thesaurus, video- and audio products of mass culture, dialogues, monologues, advertisements, etc.

These forms of cooperation are complementary (interrelated), yet, quite autonomous (independent). Socio-cultural aspect of cognitive robotics, expressed in the aforementioned forms, is closely connected with the processes taking place in people's lives. Global evolutionism is one of those processes; the global evolutionism in cognitive robotics is understood not only as awareness of various branches of science, but also as the integration of explicit and implicit socio-cultural content forms.

Additionally, improvement of methods and search for evolutionary robotics pathways can also be regarded as a certain reflection of the global evolutionism within the frame of cognitive robotics [16].

Global evolutionism tends to solve the problem of integrating sociocultural technologies into information technologies via searching the scientific / technical / engineering solution to the problem of producing and promoting cognitive robotics.

One can argue that current estimates of global evolutionism in robotics are incomplete and serve as a lame alternative [17]. Yet, the concept of global evolutionism can be considered to be methodological base, man vs. society integration process, and the robotic models per se.

The paradigm of global evolutionism predetermined the necessity to understand the processes of formation of IT society. The paradigm is regarded as man-centered one. In the context of cognitive robotized technology, the problem of understanding the world per se remains unsolved, as well as the problem of understanding society and its tech-tools – including robots.

The modern relations concerning information exchange have no final completeness, and the interactions between the social systems included in network more often have unsystematic character. The situation becomes complicated by the fact that the complexity of processing of continuously increasing volume of information resources causes the emergence of a number of the specific problems determined by the information redundancy.

It should be noted that during the present period the problem of redundancy of information acquired the special scale that dictates the necessity for further strengthening of intensity of their interactions. In general, the process of the creation of system links in the information space is caused by a large number of the phenomena of self-organization that is typical for difficult systems of a high level of uncertainty.

At the registration of structural links there is the widest range of possible trajectories of system's development that confirms the importance of the choice with the system of communications' method in the information space.

One more problem caused by the formation of a global information space is the emergence of the whole range of non-canonical, informal communications of the systems complicating the management and coordination of the processes of interaction in the network. These communications are formed owing to a combination of the mass phenomena of self-organization and imperfections of the arising network structures [18]. Thus, it is obvious that the modern situation is characterized by the registration of network structures based on active information exchanges that attracts the emergence of a number of the specific questions demanding for the solution of new innovative approaches.

Concerning social systems this principle causes the amplifying tendency to association of the systems of different complexity and the nature which are characterized by own specifics [19]. As a result, the complex structures are formed representing the global networks which are characterized by the presence of the maximum adaptation opportunities both at each participant, and at all structure.

One of the most significant manifestations of the impact of communication and information technologies on social processes the information revolution has become, promoting the formation of macro trends, based on the expansion of the range of the mutual influence and the interaction of social systems. These trends, the formation of which was largely predetermined by the enhancing processes of integration, in prospect become a kind of catalyst for globalization processes; wherein each of them to a certain extent is the manifestation of global synergy. Globalization has formed a new favorable condition for the development of social systems, at the same time creating additional factors affecting the dynamics and uncertainty of the environment. Incomplete significantly complicates the procedure for the development of optimal behavior strategies; in these circumstances, one of the most important characteristics of the social system becomes the main - the ability to quickly and efficiently adaption to the dynamics of the external environment, which is determined by the susceptibility of the system to the innovations [20, 21].

3. Summary

The significance of global evolutionism is expressed in the current formation of IT-domineering civilization, i.e., we are able to witness the growing concern about the search for optimal ways to merge humankind, society, and robotic systems into one single entity.

Sociocultural content of the global evolutionary process, which flows smoothly and continuously, predetermines the mentioned ways. It is the paradigm of global evolutionism, expressed via socio-cultural form with the help of socio-cultural approach, that presents the essence of the modern welt.

While analyzing cognitive robotized technology in accordance with individual and society, we can ponder on the internal problem examination, as well as on the issue of the robot vs. human vs. society interaction processes.

The modern relations concerning information exchange have no final completeness, and the interactions between the social systems included in network more often have unsystematic character. The situation becomes complicated by the fact that the complexity of processing of continuously increasing volume of information resources causes the emergence of a number of the specific problems determined by the information redundancy.

References

[1] Knyazeva E N, Kurdyumov S P 2012 *Zakony evolyucii i samoorganizacii slozhnyh sistem* (Moscow, Nauka)
[2] Gorbachev S, Syryamkin V 2015 *Proceedings – 2015 International Conference on Cognitive Computing and Information Processing* (CCIP)
[3] Toffler A 2014*The Third Wave* (New York, William Morrow)
[4] Drucker P 2012 *Innovation and Entrepreneurship* (New York, Collins)
[5] Makienko M A, Kvesko R B, Kornienko A A, Kvesko S B 2013 *8th International Forum on Strategic Technology (IFOST–2013)* **(2)** 692 – 693
[6] Kvesko R B, Rubanov V G, Kvesko S B, Gluchova T P 2000 *Proceedings of the 3rd International Conference on Quality, Reliability & Maintenance QRM 2000* (G J McNulty, Oxford-London)
[7] Kvesko R B, Kvesko S B, Salkova N E, Shinn T N 2005 *9th Korean-Russian International Symposium on Science & Technology KORUS 2005* 1089 – 1092
[8] Janszen F 2015 *The Age of Innovation* (London, Prentice Hall)
[9] Castells M 2010 *The Rise of the Network Society* (London, Blakwell Publishers)
[10] Nonaka I, Takeuchi H 2008 *The Knowledge-creating company: How Japanese companies create the dynamics of innovation* (Oxford, Oxford University Press)
[11] Vasendina E, Plotnikova I, Levitskaya A, Kvesko S 2016 *IOP Conference Series: Materials Science and Engineering* **110** 012070 doi:10.1088/1757-899X/110/1/012070
[12] Limoncelli T A, Chalup S R, Hogan C J 2014 *The Practice of Cloud System Administration: Designing and Operating Large Distributed Systems* (AddisonWesley Professional Publ.)
[13] Delong D, Feihi L 2010 *Diagnostika kul'turnyh bar'erov v upravlenii znaniyami. (V)Upravlenie znaniyami Hrestomatiya 2nd Edition* (S-Petersburg, Vysshaya shkola menedzhmenta)
[14] Baburin S N, Muntyan M A, Ursul A D 2015 *Globalizaciya v perspektive ustoichivogo razvitiya* (Moscow, Infra-M)
[15] Giddens E 2015 *Uskol'zayushii mir. Kak globalizaciya menyaet nashu zhizn'* (Moscow, Ves' mir)
[16] Shumilov V N, Syryamkin V I, Syryamkin M V 2015 *AIP Conf. Proc. (1688)* 040007
[17] Wallerstein I 2000 *International Sociology* (15) 6 249–265
[18] Basov N V 2008 *Issledovatel'skie strategii izucheniya social'nyh innovacii v kontekste koncepcii obshestva znaniya (V) Obshestvo znaniya: ot idei k praktike. Kollektivnaya monografiya v 3-h chastyah Chast' 1 Osnovnye kontury koncepcii obshestva znaniya* (S-Petersburg, Skifiya-print)
[19] Zak M H 2010 *Razrabotka znanievoi strategii epilog (V) Upravlenie znaniyami Hrestomatiya 2nd Edition* (S-Petersburg, Vysshaya shkola menedzhmenta)
[20] Bell D 2013 *The Coming of Post-Industrial Society A Venture in Social Forecasting* (New Basic Books)
[21] Yurchenko A *et al* 2015 *IOP Conf. Ser.: Mater. Sci. Eng.* **81** 012097

Artificial intelligence in robot control systems

A Korikov
Tomsk State University of Control Systems and Radioelectronics, Tomsk, Russia
Tomsk Polytechnic University, Tomsk, Russia
E-mail: korikov@asu.tusur.ru

Abstract. This paper analyzes modern concepts of artificial intelligence and known definitions of the term "level of intelligence". In robotics artificial intelligence system is defined as a system that works intelligently and optimally. The author proposes to use optimization methods for the design of intelligent robot control systems. The article provides the formalization of problems of robotic control system design, as a class of extremum problems with constraints. Solving these problems is rather complicated due to the high dimensionality, polymodality and a priori uncertainty. Decomposition of the extremum problems according to the method, suggested by the author, allows reducing them into a sequence of simpler problems, that can be successfully solved by modern computing technology. Several possible approaches to solving such problems are considered in the article.

1. Introduction

The The Artificial Intelligence (AI) and robotics are of interest practically to each person of our world. AI is the branch of computer science, concerned with making computers behave like humans. The term has been introduced in 1956 by John McCarthy [1]. Among many AI definitions we prefer definition given in [2]. AI is the property of the computer or of the neural network which consists in their reaction to data almost in the same way as a person reacts to information. In the monograph [2] the theory of AI is described as the science about agents who receive results of perception acts of the external environment and perform rational operations. AI defines the present and the future of the technological industry and the equipment. Robotics includes many achievements of AI [3]. Robots differ from usual technical systems by a new property, generated by synergy of mechanical, electronic and computer components of robots and AI. There is a question of assessment of new quality of the robots with AI. How to determine the level of intelligence of the robot? How the system theory and the theory of artificial intelligence answer this question? How to use the AI methods and methods of technical systems synthesis when designing intelligent robots? Let's try to answer these questions.

2. On estimation of robot intelligence

First, let's answer a question of robot intelligence degree assessment on the example of estimation of intelligence degree of natural systems. The idea of human intelligence degree assessment by tests and the notion of IQ – intelligence quotient, which is defined on the basis of these tests, was offered and developed in the beginning of 20th century. The IQ is used in many countries of the world for the achievement of the most different purposes. For the assessment of intelligence degree of AI systems various AIQ intelligence coefficients were suggested in [4], such as absolute intelligence coefficient (AAIQ), relative intelligence coefficient (RAIQ) and comparable intelligence coefficient (CAIQ). Brief distinctive characteristics of these estimates are explained in [5]. Also, the authors of [5, 6] point to the common flaw of such estimates, which consists in their subjectivity since they are based on expert assessments.

Some researchers make attempts of objective AIQ assessment. So, for instance in [6] the concept of intelligence "levels" of AI systems was introduced, which are human machine interface, event forecasting, adaptation and self-learning, events, knowledge and decision making databases, and, finally, the execution level. The term "intelligence level of AI systems" is defined in [6] in a way, that is similar to the concept of "stability of automatic control systems". System stability can be defined "in the small", "in the large" and "in general", therefore the intelligence of AI systems in [6] is also defined "in the small", "in the large" and "in general". Intelligence "in the small" means that system performance is restricted to the two bottom levels, intelligence "in the large" – to the three bottom levels; and in case of intelligence "in general" it covers all five levels. The analogy of AI system intelligence and system stability has its advantages and disadvantages. It is beyond argument that such analogy increases the objectivity of AIQ assessment, but at the same time in the theory of non-linear automatic control systems (ACS) the stability "in the small" of the ACS doesn't follow from the stability "in the large". It is necessary to correct the discussed analogy in a sense that will become obvious after reviewing the concepts of "intelligence in the small" and "intelligence in the large", suggested in [6].

The problems of intellectual robot control are on the intersection of control theory and the theory of AI. They substantially match the problems of development of the intelligent automated industrial management systems (AIMS) arising on a joint of electronics and informatics as a combination of electronic and computer devices [5]. The development of AIMS is induced by control theory, the theory of AI, system theory and system analysis. Intersection of these theories forms area I (Figure 1). The area I on Figure 1 is defined as the intelligent control [7] or control which has the property of "intelligence in the small" [6]. It is possible to define systems, implementing intelligent control, as AI systems, which has the property of intelligence "in the small" [6]. The subarea II is a part of the area I on Figure 1. This subarea defines the control, which has the property of "intelligence in the large" and corresponds to intellectual control [5, 7]. Arrows on Figure 1 illustrate the cross impact of three scientific theories while numbers 1 through 10 designate concepts and methods, which are transferred from one theory to another and create solving methodology for unformalized (semistructured) problems of complex dynamic systems control. For example, arrow 1 on Figure 1 shows that the theory of AI enriched the theory of system analysis with data processing methods, and arrow 4 points out that techniques of systems analysis are used in the theory of AI.

Concepts of adaptation and intelligence used in robotic have been approbated in control theory. Tsypkin in [8] identified three stages in the control theory: determinism, stochasticity and adaptivity. We are witnessing the development of new, fourth stage of the control theory: intelligence. The former three stages are specified by solving of formalized problems, while the latter one focuses on unformalized (semistructured) problems [8].

Systems of intelligent control (SIC) are based on five principles [5, 6]: interaction of systems with an external environment; system openness; forecasting external environment and internal behavior changes; layered system architecture; system survival capability in the case of a failure of communication with the highest levels of system structure.

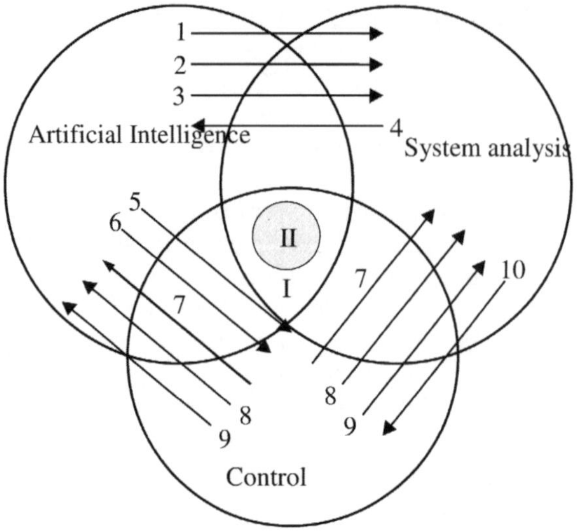

1. Data processing	6. Learning
2. Formal linguistics	7. Adaptive control optimization
3. Sets of heuristics	8. Dynamic feedback
4. System analysis	9. Control system dynamics
5. Computer memory	10. Coordinating control

Figure 1. Interaction of basic theories of intellectual robotics.

These principles of SIC design allow defining the terms "intelligence in the small" and "intelligence in the large" [6]:

1. SIC are referred to as intelligent-in-the-large systems if they are organized and perform in full accordance with five principles given above.
2. SIC are referred to as intelligent-in-the-small systems if they aren't organized and don't perform in full accordance with five principles given above, but use knowledge databases, automatic control or behavioral models in order to overcome vagueness of input data.

Examples of both types of SIC are described in [5 – 7]. From the definitions given above it follows that intelligent-in-the-large SIC is also intelligent-in-the-small. These concepts correspond to the known concepts of "weak-AI" and "strong-AI" [2]. The concept of "intelligence in general" remains ill-defined as in the theory of AI there is no predominating point of view on an interpretation of this level of intelligence yet.

One of the founders of the system theory William Ross Ashby [9] enunciated the law of requisite variety which considers limit implementability of system. This law for control systems can be formulated as follows: controlling subsystem V_{cs} should be more diverse (complex) than controlled subsystem (controlled object V_{co}): $V_{cs} > V_{co}$. Therefore, the controlling subsystem of SIC should have the variety (complexity) of methods, models, algorithms and controls larger than complexity of an object. The law of requisite variety has also the second corollary which is defined as follows: if variety at top levels of SIC structure increases, then variety at lower levels is restricted and, vice versa, increase in variety at the bottom levels disintegrates top levels. This corollary is also referred to as the law of hierarchical compensations (Sedov law).

The coefficient of intelligence (AIQ for SIC) is proportional to variety (complexity) of solvable tasks (tests, problems). This statement and two corollaries of the law of requisite variety lead to the following consequences for the systems of intelligent robot control (SIRC):

- SIRC with large variety of the software has a high level of intelligence (high coefficient of intelligence AIQ). Therefore, the architecture of such SIRC is complex, multilayered and has multiple loops.

- SIRC with large variety at top levels and, respectively, with a high level of intelligence at these levels should have limited variety at the bottom levels (restriction on degree of intelligence). This restriction at the bottom levels provides automatism of execution, high accuracy and high-speed performance. Therefore, increase in intelligence "in the large" restricts intelligence "in the small".

3. Multilevel control systems with optimization

The robotic technology uses many mathematical methods of a research. Methods of optimization are especially attractive [5, 7, 10, 11]. Systems of control with optimization (SCO) use these methods. Multilevel control systems with optimization (MSCO) can be described by the block diagram which is submitted in Figure 2.

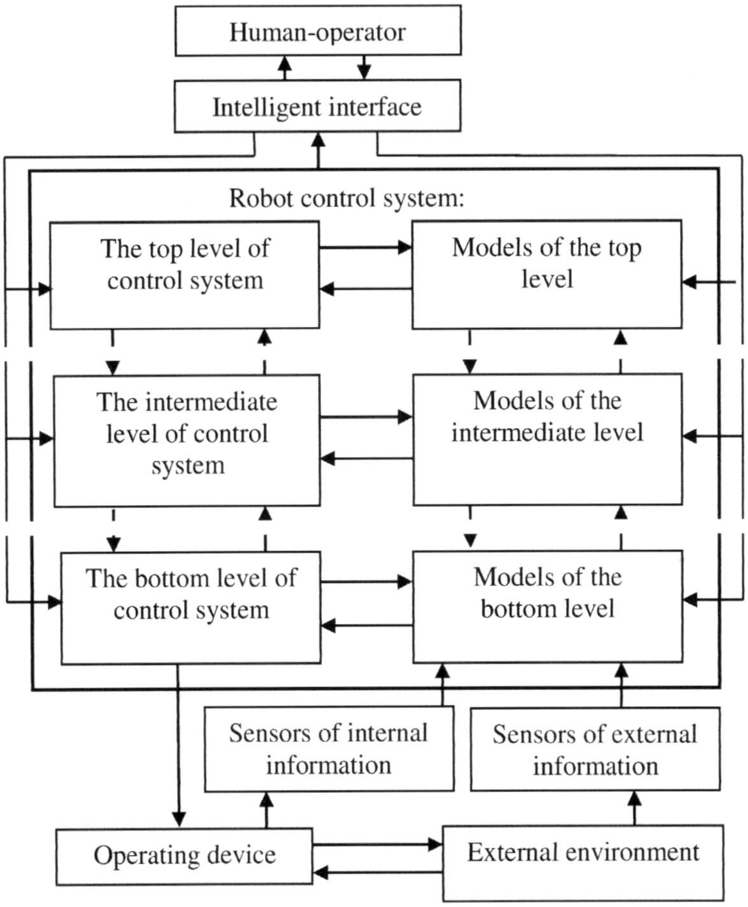

Figure 2. Structure of MCSO of the robot.

Functioning of the MCSO is based on the extremal principle which lie in the fact that some functional of J (x) from state vector x (motion trajectory or structures) the MCSO of the robot accepts extremal value. Overall performance the OCS of the robot is defined by the accuracy of determination of extremal value (minimum or maximum) x^* a state vector of this functional: x^*= argextr J (x). Purpose the MCSO consists in determination of state variables – arguments of a functional of J(x) which provide point of extremum (minimum or maximum) this functional. Our approach we will consider on examples three-level, two-level and single-level the MCSO of robots. Decision-making at all levels the MCSO of robots is based on the solution of extremum problems of different types. Problem description of research of MCSO is offered in our article [11]. Formalization of this problem description consists in the job of three sets: X_E - a set of states of the external environment (EE) which interact with the robot; X_O - set of states of controlled object (CO); X_M - set of states of the MCSO models. Therefore general case of the state space of the researched system represents the Cartesian product of the listed sets: $X= X_E \times X_O \times X_M$. Each point x of space of X is the characteristic of a status of the robot, his OCS and interactions of the robot with an external environment. Functioning of the robot represents movement of a point x in space of X, that is means its transition in space of X of one status to other status on some path:

$$\mathbf{x}(\mathbf{c},\mathbf{u},t)=\left[\mathbf{x}_e(\mathbf{c},t),\mathbf{x}_o(\mathbf{u},t),\mathbf{x}_m(\mathbf{x}_o,t)\right].$$

The vector of the controlling signals is designated as u, and the vector of parameters of models of system of control of the robot is designated as c. The vector of parameters of models c consists of three components: c = (c_t,c_m,c_b). That is, components of vector of c are vectors of parameters of models of the top level of c_t, the intermediate level of c_m and the bottom level of c_b of system of control of the robot. The state vector the MCSO of the robot consists of vectorial components: x=(x_e, x_o, x_m); $x_e \in X_E$, $x_o \in X_O$, $x_m \in X_M$.

Efficiency of functioning the MCSO of the robot is estimated by the job of target functions of $F_i(i=1,2,...,n)$ and restrictions of $H_j(j=1,2,...,m)$ on a set of trajectories of the MCSO of the robot in the state space X. The control system of the robot solves the following extremum problem:

$$J\left\{F_1\left[\mathbf{x}(\mathbf{c},\mathbf{u},t)\right],...,F_n\left[\mathbf{x}(\mathbf{c},\mathbf{u},t)\right]\right\} \to \underset{\mathbf{x},\mathbf{c},\mathbf{u}}{extr}, \quad (1)$$

$$H_j\left[\mathbf{x}(\mathbf{c},\mathbf{u},t)\right] \geq 0, \quad j=1,2,...,m. \quad (2)$$

This task is the task on search of an extremum of a functional of J in case of restrictions of H_j (j = 1, 2,…, n), that is it is required to define c, u and to find x which meet a condition (1) in case of restrictions (2). The decision of similar tasks sophisticates dimensionality of the task, its polymodality, prior uncertainty, etc. Our papers [6, 11] offer a technique of the decision of such tasks. Decomposition of the task (1) by this technique supposes its partition on tasks of three levels which correspond to three MCSO levels in Figure 2. The functional of J is set by the human-operator. The top level the MCSO defines components of a vector c_t^* and creates the sequence of local goals of

$$F_i\left[\mathbf{x}(\mathbf{c}_t^*,\mathbf{c}_m,\mathbf{c}_b,\mathbf{u},t)\right], \quad i=1,2,...,n,$$

as the solution of the following extremum problem:

$$J\left\{F_1\left[\mathbf{x}(\mathbf{c},\mathbf{u},t)\right],...,F_n\left[\mathbf{x}(\mathbf{c},\mathbf{u},t)\right]\right\} \to \underset{\mathbf{c}_t,\mathbf{x}(\mathbf{c}_t,t/\mathbf{c}_m,\mathbf{c}_b,\mathbf{u})}{extr} \quad (3)$$

in case of restrictions (2).

The intermediate level of MCSO solves extremum problems:

$$F_i\left[\mathbf{x}(\mathbf{c}_t^*,\mathbf{c}_m,\mathbf{u},t)\right] \to \underset{\mathbf{c}_m,\mathbf{x}(\mathbf{c}_m,t/\mathbf{c}_t^*,\mathbf{c}_b,\mathbf{u})}{extr} \quad (4)$$

in case of restrictions (2).

The bottom level of MCSO solves tasks of coordinate or parameter optimization of CO. Coordinate optimization of CO defines the combination of input coordinates (variables) CO providing x_o extremum – one of components of a vector $x_O \in X_O$ which characterizes quality of operation of CO. Parameter optimization of CO defines a set of parameter values of CO which provides an extremum of criterion of the quality Ф created at the lowest level of MCSO. Thus, the lowest level of MCSO solves one of the following extremum problems:

$$x_o\left[\mathbf{x}_m\left(\mathbf{c}_t^*,\mathbf{c}_m^*,\mathbf{c}_b,t\right),\mathbf{u}\right] \to \underset{\mathbf{c}_b,\mathbf{u}}{extr}, \quad (5)$$

$$\Phi\left[\mathbf{x}_o(\mathbf{u},t),\mathbf{x}\left(\mathbf{c}_t^*,\mathbf{c}_m^*,\mathbf{c}_b,t\right),\mathbf{u}\right]\to \underset{\mathbf{c}_b,\mathbf{u}}{extr} \quad (6)$$

in case of restrictions (2).

Thus, the extremum problem (1) is reorganized into set of simpler extremum problems (3) - (6). These tasks make a clear physical sense in case of control of specific CO. The top levels of MCSO use methods of an artificial intelligence, and lower – methods of automatic control. Let's show it on examples of production control (PC) and the transport robot (TR).

The top level of MCSO of PC solves the problem (3). The decision of the task (3) is presented in the form of the law of control of PC, initial and boundary conditions. The intermediate level of MCSO of PC solves the problem (4), that is realizes setup of parameters in the selected law of control of PC. The bottom level of OCS of PC solves the task (5) or the task (6). The circuit of control consists of the lowest level of OCS, CO and sensors of information on a status of CO and represents the known systems of extremal control (extremal ACS) [6, 7, 10-12].

The same transparent physical sense of the task (3)–(6) has in case of control of TR: The OCS of TR is intellectual system of navigation and control of TR (SNCTR). The top level of SNCTR solves the problem (3), plans a global route of movement of TR in the region and defines location of TR in the coordinate system connected to this region. The intermediate level of SNCTR solves the problem of a kind (4), orients TR on locality of relative to navigation reference points and plans a local route of its movement. The SNCTR bottom level solves problems (5) or (6) and exercises optimum driving of TR on locality.

4. Conclusion

Research in the field of AI is currently being conducted intensively and very diversely. In our work an attempt is made to answer the question: is it possible to determine the measure of intellectuality of robotic systems? The criteria of intellectuality of robotic systems are considered. Estimates of the intelligence of robotic systems are associated with their complexity and diversity. In the case of robotic systems, the definition of SIC as systems that act rationally and optimally is relevant [2].

References

[1] McCarthy J 1989 *Artificial Intelligence, Logic and Formalizing Common Sense. In Richmond Thomason, editor, Philosophical Logic and Artificial Intelligence* (Klüver Academic)
[2] Russell S, Norvig P 2016 *Artificial Intelligence: A Modern Approach* . Third Edition. (Pearson Education, Limited)
[3] Bekey G 2008 *Robotics: State of the Art and Future Challenges* (Imperial College Press)
[4] Litvinov V, Oksanich I 2008 *Mathematical machines i systems* **(2)** 100–105
[5] Korikov A 2011*Mekhatronika, avtomatizatsiia, upravlenie* **(5)** 2–8
[6] Lokhin V, Zakharov V 2001 *Mekhatronika* **(2)** 27–35
[7] Vasil'ev S, Zharkov A, Fedosov E, Fedunov B 2001 *Intellektny management of dynamic systems.* (Moscow, Fizmatlit)
[8] Tsypkin Ia 1982 *Vestnik AN SSSR* (9) 116–121
[9] Ashby W 1963. *An Introduction to cybernetics* (Methuen)
[10] Goritov A, Korikov A 2001 *Automation and Remote Control* **62** (7) 82–90
[11] Korikov A 2014 *Scientific Bulletin of the NSTU* **(2)** 18–26
[12] Serikov M Z 2017 *Proceedings – 2016 11th International Forum on Strategic Technology, IFOST 2016* 510

A research of the propagation of LoRa signals at 433 and 868 MHz in difficult urban conditions

V Eremin, A Borisov
Altai State Technical University named by I.I. Polzunova, Barnaul, Russia
E-mail: boralp@mail.ru

Abstract. The article examines the issue of investigating the propagation of LoRa signals at 433 MHz and 868 MHz in difficult urban conditions. A theoretical calculation was made, which showed that the signal propagation distance is about 1.8 km for the module with a frequency of 433 MHz and about 915 m for the 868 MHz frequency. Experimental studies show that the transmission distance of the 868 MHz module is approximately 660 m, module 433 MHz – 730 m. The discrepancy is due to the influence of interference, which is always present in the modern city, as well as with complex and diverse buildings, which is not taken into account in the parameter SOM when calculating the transmission range.

1 Introduction

The direction of the Internet of things in the last decade is one of the promising areas for the development of communication and information technologies. This is due to the increasing information needs of people and the growing complexity of modern information systems [1].

The Internet of things is a network of physical devices, vehicles, structures and other objects with built-in electronics, software, sensors and Internet connection. All this allows you to collect and share data, which in turn excludes from the part of operations and actions human participation in social and economic processes [2].

Thus, the concept of a network of computing devices equipped with built-in technologies for transmitting and receiving data between themselves and with the environment, represents the organization of such networks as a phenomenon capable of transforming social and economic processes, without human intervention [3, 4].

All these technologies, which allow a qualitative increase in the standard of living, are subdivided into methods of transmission and interaction (ZigBee, 6LoWPAN, WirelessHart, LPWAN), power supply methods for devices (solar batteries, windmills, etc.). All this is based on different physical phenomena.

LoRa ™ (Long Range) is a technology and the same modulation method. The LoRa modulation method is patented by Semtech, based on the technique of spreading and the variation of linear frequency modulation, in which the data is encoded by wideband pulses at a frequency that increases or decreases over a certain time interval. This solution, in contrast to the direct spread spectrum technology, makes the receiver stable to frequency deviations from the nominal value and simplifies the requirements for a clock generator, which makes it possible to use inexpensive quartz resonators. LoRa uses direct error correction, operates in the sub-Higahertz frequency range [5–7].

The Lora wireless system uses non-licensed frequencies that are available around the world. The most widely used frequencies are 915 MHz, 868 MHz, 433 MHz. Low power consumption and high stability of communication at a great distance are provided due to high sensitivity of –148 dBm [8].

The aim of the work is an experimental study of the propagation of the LoRa signal in complex urban conditions.

The tasks of the work include:

- develop a software and hardware system that allows data transfer using the LoRa standard;
- conduct research on data transmission at a frequency of 433 Hz and 868 MHz;
- compare the results obtained during the study.

2 Principle of operation

To solve the set tasks, a laboratory complex using the LoRa standard was developed. It should consist of a server device and several client devices. The hardware can be found in the form of ready-made development kits, but they do not fully allow you to study this technology in the form in which you can buy them, so they need to be finalized. As a consequence, there is a need to create your own devices. As a server device is the hardware platform Arduino pro mini - a programmable board based on the microcontroller Atmega328, with a radio module Lora and a screen 0.43 '. The connection to the computer is via the PL2303 chip. The client version of the device must contain a GPS module, in addition to a motherboard based on the Atmega328 microcontroller and the Lora module (Figure 1).

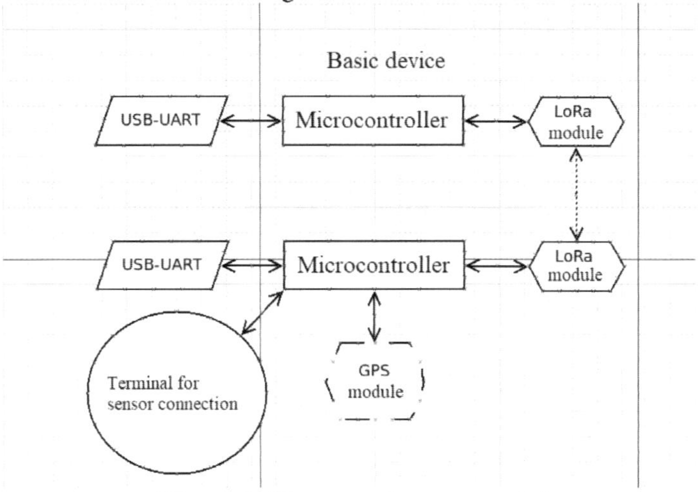

Figure 1. Diagram of system modules.

Figure 2 shows a block diagram of the operation of the device.

The principle of the device is as follows [9]:

• power is supplied and the server and client device is started;

• the client device receives data from the GPS module and transmits them via the radio channel to the server one;

• the server device waits for data from the client device;

• after receiving them, sends coordinates through the serial port to the application running on the computer. The level of the received signal is also calculated;

• the application running on the computer after receiving the data, processes them and draws the corresponding client device on the OpenStreetMap card;

• after receiving the data, the server device reads the values from the control unit, and then it is checked whether the received signal level indicator of the set threshold has exceeded the control unit.

If the threshold is exceeded, an audible alert is made.

The tasks of the main device are:

• indication of the current signal level in dBm;

• transmission of the exact coordinates of the auxiliary device to determine the distance between the devices;

• receiving coordinates in an encrypted form;
• Record the received data on the media. (when connecting to a PC).
The task of the auxiliary device is reduced to:
• Receiving coordinates from GPS module and sensor (optional);
• transmission of received data in encrypted form.
Images of the server and client devices are shown in Figures 3 and 4.

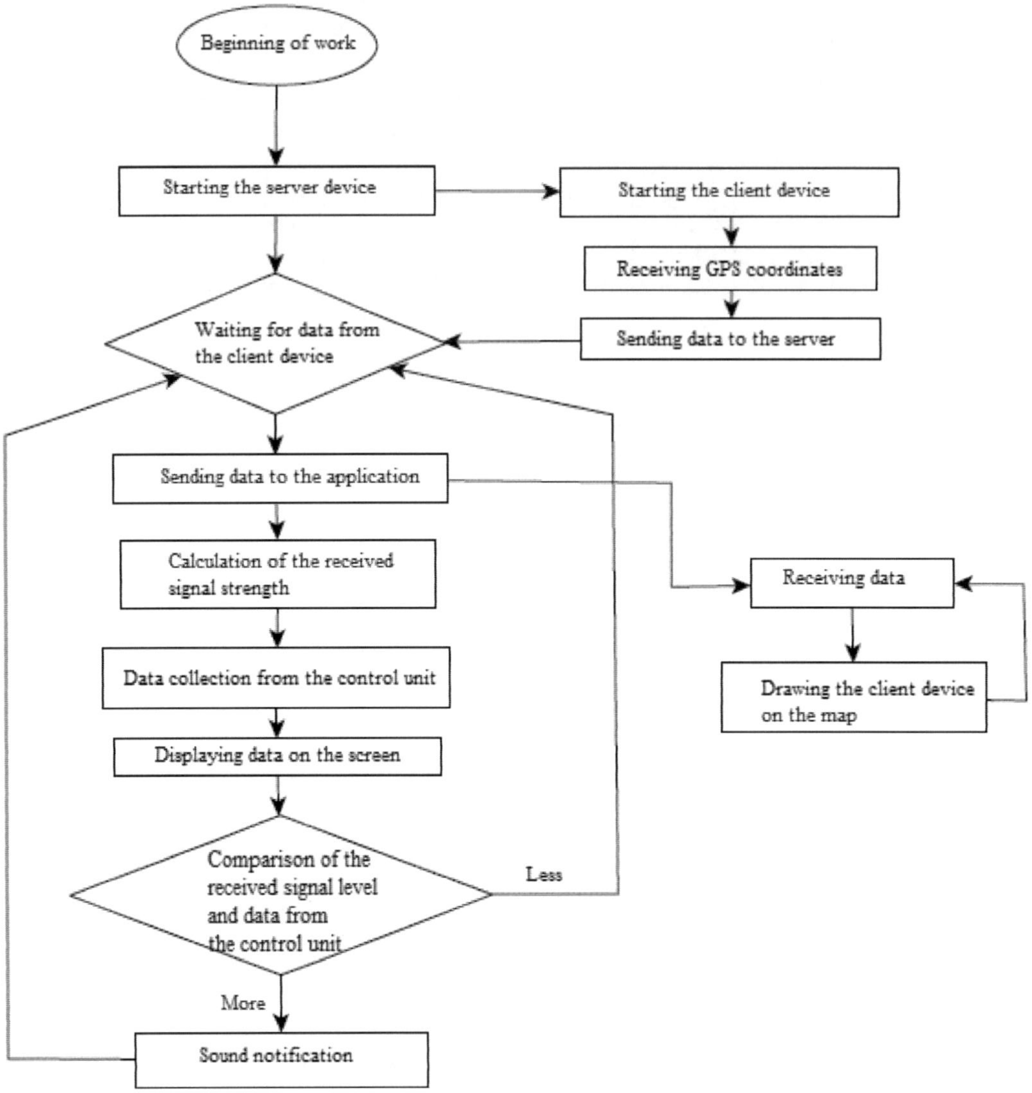

Figure 2. Block diagram of the device operation.

Figure 3. Prototype of the server device.

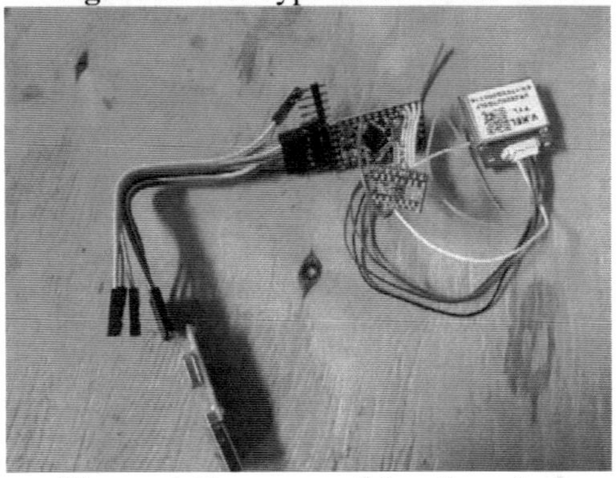

Figure 4. Prototype of the client device.

3 Results of device testing

To study the propagation of the signal, we calculate the transmission range of information. The formula for calculating the transmission range FSL is determined by the total gain of the system. The total gain of the system is considered as follows:

$$Y = P_t + G_t + G_r - P_{min} - L_t - L_r ,$$

where Y is the total gain of the system (dB), Pt is the transmit power (dBm), G_t is the transmit antenna gain (dBi), G_r is the receive antenna gain (dBi), P_{min} is the receiver sensitivity at this speed (dBm), L_t - loss of signal in the coaxial cable and connectors of the transmission path (dB), L_r - loss of signal in the coaxial cable and the connectors of the receiving path (dB).

FSL is calculated by the formula:

$$FSL = Y - SOM ,$$

where *SOM* (System Operating Margin) - reserve in the energy of radio communication (dB).

The SOM parameter is usually taken equal to 10 dB. It is believed that a 10-decibel margin for amplification is sufficient for engineering calculation.

As a result, we obtain the communication range formula:

$$D = 10^{\left(\frac{FSL}{20} - \frac{33}{20} - lgF\right)},$$

where D is the communication range in meters.

For the study, a LoRa data transmission module based on the RFM96W chip was chosen. The transmitter's declared transmitter power is 20 dBm at 100 mW, the sensitivity is -148 dBm.

The transmission range was calculated for the case with a standard antenna without gain.

By default, the transmitter power is set to 100%. Declared by the manufacturer, the transmission distance in urban conditions is 1–2 km.

When calculating the signal attenuation, it turns out that with a standard antenna FSL = 151 dB.

Then the transmission distance is: for a module with a frequency of 433 MHz - about 1.8 km, and for a frequency of 868 MHz - 915 m [10].

To confirm the calculated data, a full-scale experiment was carried out. The general view of the hardware and software complex is shown in Figure 5.

Figure 5. Schematic diagram of the laboratory complex.

The results of the experiment are shown in Figure 6.

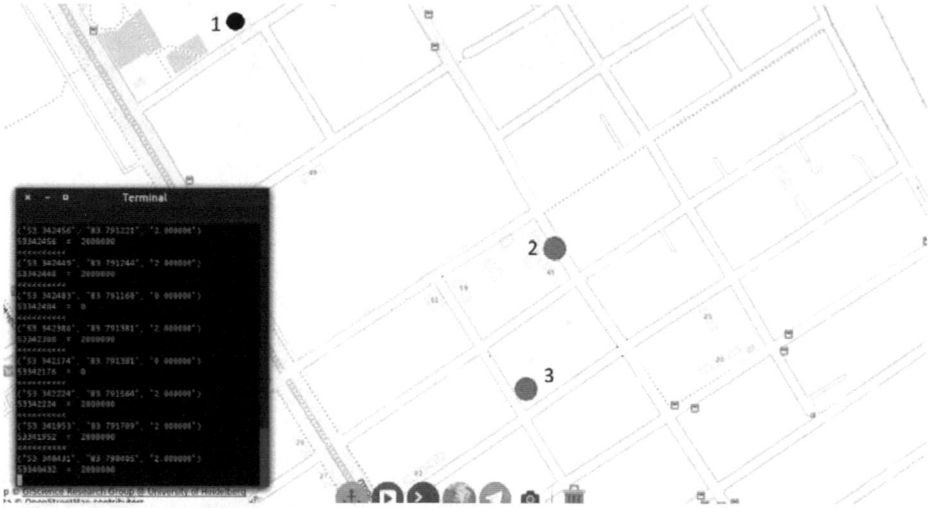

Figure 6. Results of the experiment: 1 – Transmitter, 2 – Receiver 868 MHz, 3 – Receiver 433 MHz.

The transmitter was at a level of 10 meters from the ground. With a further increase in range, the data is unstable, until the signal is completely lost.

The distance traveled is shown in Figure 7.

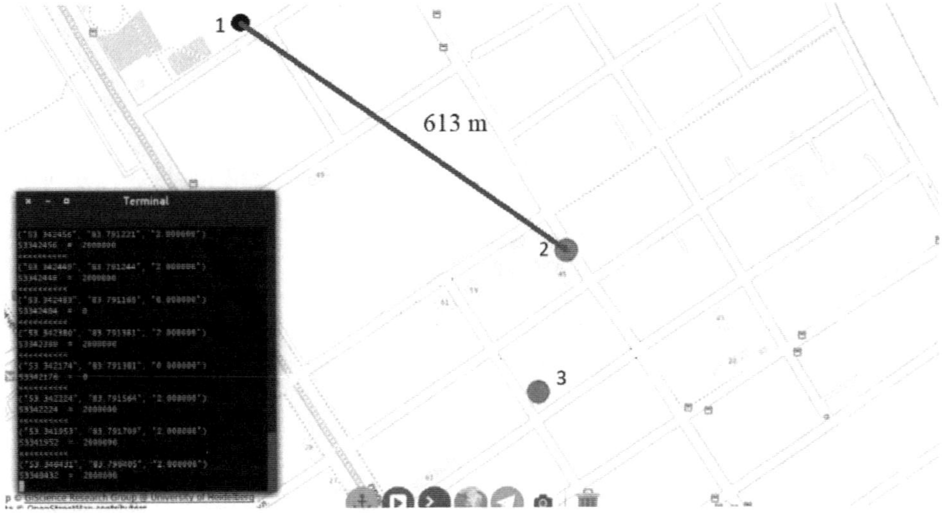

a)

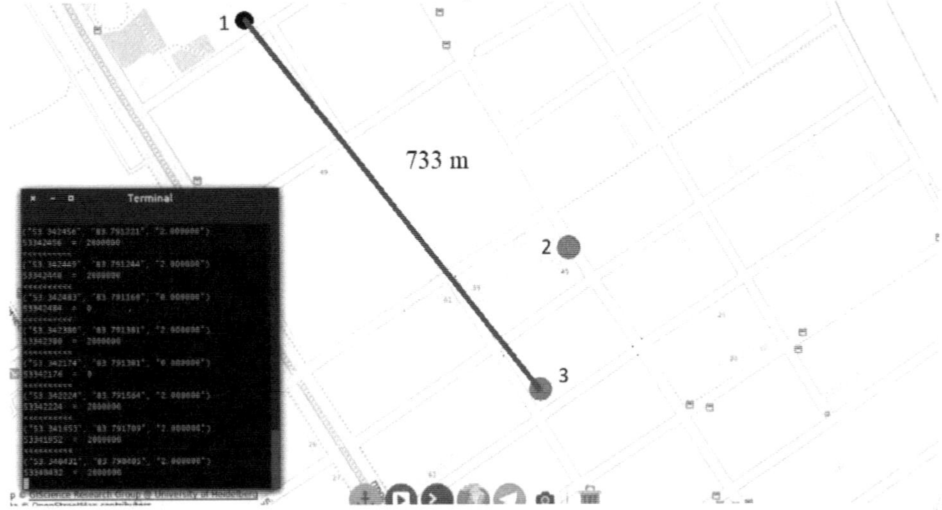

b)

Figure 7. Distance traveled by receivers: a) the distance traveled by the 868 MHz receiver, b) the distance traveled by the 433 MHz receiver

These figures show that the transmission distance of the 868 MHz module is approximately 660 m, the module 433 MHz – 730 m.

This is because the SOM parameter does not take into account interference from devices operating at 433 and 868 MHz, for example, car alarms, and does not fully take into account the complexity and frequency of urban development.

4 Conclusion

When comparing the calculated values of the transmission range in the LoRa wireless channel at 433 and 868 MHz with the experiment performed, it can be concluded that there is a significant discrepancy between the theoretical calculation and the practically obtained data. This is due to the influence of interference, which is always present in a modern city, and also with a complex and diverse build-up, which is not taken into account in the SOM parameter when calculating the transmission range.

References

[1] Samuel G. *Internet of things: The future is here* (Alpina Pablisher, Moscow, 2016)
[2] Zaramensky E P, Artemyev I E 2016 *Internet of things: Research and field of application* (Moscow, Infra-M)
[3] Kumaritova D L, Kirichek R V 2016 *Information Technologies and Telecommunications* **4** 33
[4] Daud Sh *et al* 2018 *IOP Conf. Ser.: Mater. Sci. Eng.* **318** 012053 doi:10.1088/1757-899X/318/1/012053
[5] Azmi N *et al* 2018 *IOP Conf. Ser.: Mater. Sci. Eng.* 318 012051 doi: 10.1088/1757-899X/318/1/012051
[6] Jing Y and Lu H 2013 *Journal of Semiconductors* **34** (10) 105006 doi: 10.1088/1674-4926/34/10/105006
[7] Krivchenko T 2007 *Wireless technologies* **1** 38
[8] Demko A I, Radomsky S A 2014 *T-Comm - Telecommunications and Transport* **3** 19
[9] Eremin V B, Borisov A P 2017 Certificate of state registration of the computer program No. 20107660998, 03.08.17, publ. 10/02/17
[10] Eremin V B, Borisov A P 2016 *International Scientific and Practical Conference* **4** 100

Mathematical modeling of tomographic scanning of cylindrically shaped test objects

B I Kapranov, G V Vavilova, A V Volchkova and I S Kuznetsova
National Research Tomsk Polytechnic University, Tomsk, Russia
E-mail: wgw@tpu.ru

Abstract. The paper formulates mathematical relationships that describe the length of the radiation absorption band in the test object for the first generation tomographic scan scheme. A cylindrically shaped test object containing an arbitrary number of standard circular irregularities is used to perform mathematical modeling. The obtained mathematical relationships are corrected with respect to chemical composition and density of the test object material. The equations are derived to calculate the resulting attenuation radiation from cobalt-60 isotope when passing through the test object. An algorithm to calculate the radiation flux intensity is provided. The presented graphs describe the dependence of the change in the γ-quantum flux intensity on the change in the radiation source position and the scanning angle of the test object.

1. Introduction

The analysis of the internal structure of the test object is required to solve various problems of non-destructive testing [1–6]. X-ray tomography uses X-rays. The initial data for the analysis of the object internal structure are projections (X-ray patterns). An X-ray pattern is a two-dimensional shadow image. The structural elements in the image overlap, which does not allow examination and comparison of individual local fragments [7–8] in different parameters. However, if an X-ray pattern is produced from different directions, more data on the object internal structure can be obtained [9, 10].

This problem can be solved by means of X-ray computed tomography [11, 12]. The X-ray computed tomography measures spatial distribution of a specific physical quantity from different directions and computes the images free of interference caused by overlapped structures [7, 9, 13].

The main components of the X-ray tomograph are a tube, a detector and a rotary mechanism [11, 13–15]. To obtain a tomographic image, the test object is placed between the tube and the detector. During scanning, the test object is rotated by 360° [11, 16]. X-rays interact with the test object material and form its shadow image on the detector. Thus, a set of shadow images of the scanned object is obtained, and the array of the images is used for mathematical reconstruction of the object cross section [17].

This paper describes a program for modeling tomographic scanning in first-generation geometry, i.e. the source is regarded as a point γ-ray source.

In survey radiography, a relative X-ray intensity distribution is recorded. The radiation intensity of the detector decreases exponentially as the scanned sample thickness increases [18].

2. Geometry of test object scanning

To write algorithms and programs for collecting projection data and mathematical reconstruction, tomography uses test objects that have a set of standard inhomogeneities with designed geometric dimensions located within the reconstruction zone [19–21].

Figure 1*a* shows the section of a cylindrically shaped test object with the radius r0 base.

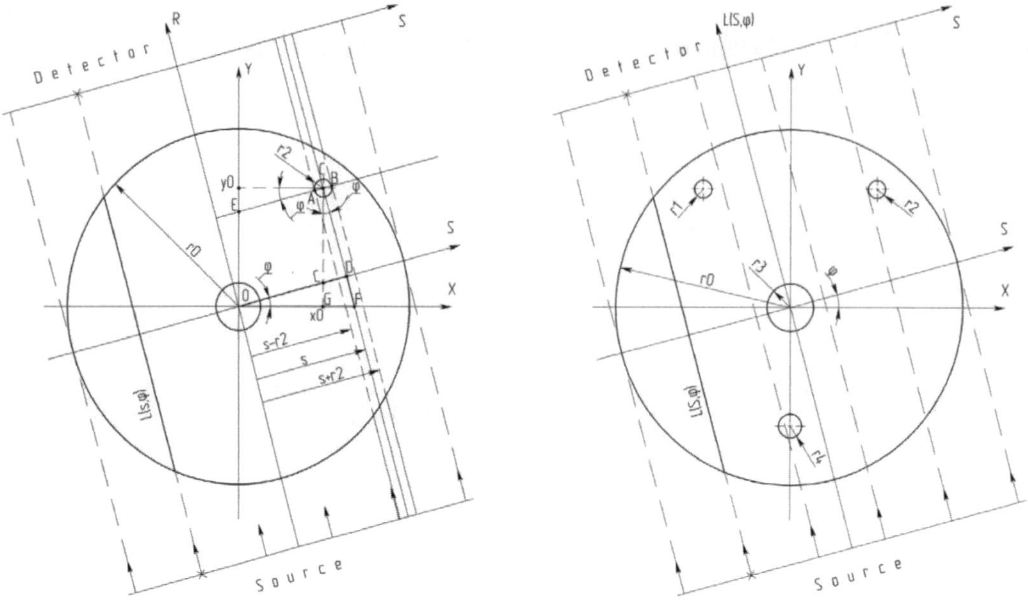

a) b)

Figure 1. Geometry of the test object cross section, where X and Y are coordinate axes in the Cartesian coordinate system; x and y are distances to an arbitrary point in the Cartesian coordinate system; S and R are coordinate axes in the scanning system; s and r are distances to an arbitrary point in the scanning system; φ is the angle of the scanning system rotation with respect to the Cartesian coordinate system.

The test object center coincides with the reconstruction zone center (t.0). The hole center with radius *r2* lies in the XY plane at point (x0, y0).

For computational convenience, turn from the Cartesian coordinate system to the polar coordinate system. In this case, the coordinate axes of the Cartesian coordinate system (X and Y) are replaced by the coordinate axes in the scanning system (*S* and *R*) through the system rotation by angle φ. Then the hole center with radius *r2* in the scanning system (*SR* plane) is located at point (s, r), the value of which depends on rotation angle φ.

The source radiation absorption intensity largely depends on the test object thickness. In the presented geometry (see Figure 1*a*), the test object thickness changes as it moves along the S axis, i.e. the test object thickness is equal to the circle chord length L(*s*).

For a circle with radius *r0* (in case of holes inside it), the chord length is calculated by equation [18]:

$$L(s) = 2r0 \cdot \sqrt{1 - \left(\frac{s}{r0}\right)^2} \tag{1}$$

where *r0* is the test object radius; *s* is the distance to an arbitrary point in the scanning system through which the chord passes.

The holes in the test object decrease the test object thickness, i.e. the chord length reduces. To find the resulting chord length in case of a hole inside the object, consider the hole with radius *r2* in Figure 1*a*.

To find the *OD* distance (the position of the center in the hole with radius *r2* on the S axis), one must know the length of the segment *OF* (projection of the position of the center in the hole with radius *r2* on the X axis). To find the length, consider the Δ*DOF* triangle. Since $\frac{OD}{OF} = \cos\varphi$, then

$$OD = OF \cdot \cos\varphi . \quad (2)$$

The *OF* segment consists of two segments OG and GF. According to Figure 1*a*, the segment *OG* = *x0*, hence

$$OF = x0 + GF . \quad (3)$$

According to the theorem on the equality of angles with mutually perpendicular sides, ∠*GCF* is equal to angle φ (see Figure 1*a*). In this case $\mathrm{tg}\,\varphi = \frac{GF}{CG} = \frac{GF}{y0}$. Hence

$$GF = y0 \cdot \mathrm{tg}\,\varphi . \quad (4)$$

Substitute expression (4) into (3) to have the expression for the segment *OF* in the form [9]:

$$OF = x0 + y0 \cdot tg\varphi . \quad (5)$$

Substitute the obtained data (5) into (2), and it turns out that

$$OD = OF \cdot \cos\varphi = (x0 + y0 \cdot \mathrm{tg}\,\varphi) \cdot \cos\varphi = x0 \cdot \cos\varphi + y0 \cdot \sin\varphi .$$

Consequently, the coordinates of the hole center in the scanning system depend on angle φ and can be calculated by the equation:

$$s(\varphi) = x0 \cdot \cos\varphi + y0 \cdot \sin\varphi . \quad (6)$$

The hole chord length $l(\varphi, s)$ depends on angle φ and on the distance to the hole center *s*. For the given angle φ, the chord length *l* of hole *r2* (Figure 1*a*) takes the maximum value equal to $2 \cdot r2$ if $s = x0 \cdot \cos\varphi + y0 \cdot \sin\varphi$ (in the hole center), and it is of minimum value equal to zero if $s = [x0 \cdot \cos\varphi + y0 \cdot \sin\varphi] - r2$ and $s = [x0 \cdot \cos\varphi + y0 \cdot \sin\varphi] + r2$ (at the hole edges) [9, 13].

Similarly to (1), the hole chord length is described by the expression

$$l(s,\varphi) = 2 \cdot r2 \cdot \sqrt{1 - \left[\frac{s - (x0 \cdot \cos\varphi + y0 \cdot \sin\varphi)}{r2^2}\right]^2} , \quad (7)$$

where *s* is distances to an arbitrary point in the scanning system; *r2* is the radius of the hole in question.

The presence of a hole decreases the chord length, which can be calculated by the equation:

$$L1(s,\varphi) = L(s,\varphi) - l(s,\varphi) . \quad (8)$$

3. Calculation of radiation intensity attenuation

In survey radiography, relative X-ray intensity distribution is recorded. To calculate the attenuation coefficient, it is necessary to measure both the primary radiation intensity (from the source) and the detector radiation intensity, which decreases exponentially as the thickness of the scanned object increases [18, 22].

The resulting radiation attenuation along the ray path through the object is determined by the dependence

$$P(s,\varphi) = P_0 e^{-\int_0^L \mu[l(s,\varphi)]dl} , \quad (9)$$

where P_0 is the radiation source intensity; $P(s, \varphi)$ is the radiation intensity of the detector; $\mu[l(s, \varphi)]$ is linear radiation attenuation coefficient at point l along ray $L(s, \varphi)$.

Linear attenuation coefficient μ depends on medium density ρ and mass attenuation coefficient μ_ρ

$$\mu = \mu_\rho \cdot \rho . \tag{10}$$

Mass attenuation coefficient μ_ρ is determined by chemical composition of the substance, photo absorption cross sections τ, Compton scattering σ, and pairing æ for each of the constituent elements, which values depend on radiation energy E. The total mass attenuation coefficient is calculated by the equation $\mu_\rho = \sum_i \omega_i \cdot \mu_\rho^i$.

where ω_i, % is percentage content of the i-th element in the substance of the test object.

For the test object without any holes inside, but taking into account the attenuation coefficient, equation (1) can be written in the form

$$L(s) = 2\mu_0 \cdot r0\sqrt{1-\left(\frac{s}{r0}\right)^2}, \tag{11}$$

where μ_0 is the linear coefficient of radiation attenuation of the test object material.

The γ-quantum flux intensity can be also affected by holes inside the test object. Taking into account the linear radiation attenuation coefficient of the hole materials, equation (7) takes the form (similar to (1) and (11))

$$l(s,\varphi,\mu_i) = 2 \cdot r_i \cdot \mu_i\sqrt{1-\left[\frac{s-(x_i \cdot \cos\varphi + y_i \cdot \sin\varphi)}{r_i^2}\right]^2}, \tag{12}$$

where s, is distances to an arbitrary point in the scanning system; r_i is the radius of the hole in question.

The presence of a hole decreases the length of the ray along which radiation propagates. In the presence of n holes, taking into account the attenuation coefficients of each hole, the total decrease in the ray length can be calculated by the equation similar to (8):

$$L1(s,\varphi,R_i,\mu_i) = L0(s,\varphi,\mu_0) - \sum_{i=1}^{n} l(s,\varphi,\mu_i), \tag{13}$$

where n is the number of holes; R_{0i} is the position of the i-th hole center; μ_i is attenuation coefficient of each hole; φ is projection angle (from 0 to 2π); r_i is the radius of the i-th hole.

During transition from the Cartesian coordinate system to the polar one, coordinates of any point (x_i, y_i) can be converted into value R. According to the Pythagorean theorem, it can be calculated by the equation $R_i = \sqrt{x_i^2 + y_i^2}$.

Then expression (7) can be written in the form $l(s,\varphi,\mu_i) = 2 \cdot r_i \cdot \mu_i\sqrt{1-\left[\frac{s-R_i \cdot \cos\varphi}{r_i^2}\right]^2}$, where s, is distances to an arbitrary point in the scanning system; r_i is the radius of the hole in question; R_i is the distance to the hole center.

4. Geometry of the test object scanning

Cylindrical test objects with a set of standard inhomogeneities with known geometric dimensions located within the reconstruction zone are used to develop algorithms and programs for collecting projection data in the reconstruction tomography [13, 23].

The studied test object is a cylinder with a base radius *r*0 = 30 cm with 4 round holes. The radii and coordinates of the centers are equal to: *r*1 = 1.5 cm, point *A*1 (-15; 20); *r*2 = 1.5 cm, point *A*2 (15; 20); *r*3 = 4 cm, point *A*3 (0; 0); *r*4 =2 cm, and point *A*4 (0; 20) (Figure 1*b*).

The positions of the hole centers in the new coordinate system are found by the Pythagorean theorem (R_1 = 25 cm, R_2 = 25 cm, R_3 = 0 cm, and R_4 = 20 cm). The displacement angle of each hole relative to the origin is equal to: $\theta_1 = \pi - \text{arctg}\left(\frac{20}{15}\right)$, $\theta_2 = \text{arctg}\left(\frac{20}{15}\right)$, $\theta_3 = 0$, and $\theta_4 = \frac{3\pi}{2}$.

The chemical composition of the test-object material can be described by the formula $Ba(NO_3)_2$+ $C_6H_{10}O_5$. The composition density is ρ = 1.8 g/cm^3. Since the paper considers only one arbitrary section of the object, the object height is not of interest to us.

Isotope cobalt-60 (Co^{60}) is used as a radiation source, which γ-radiation comprises two 1.173 MeV and 1.332 MeV lines located side by side. Therefore, γ-radiation from Co^{60} is considered to be monochromatic at energy E_0=1.25 MeV [11].

Table 1 summarizes the values for mass coefficients of radiation attenuation and the percentage content for each chemical element found in the test object material at energy of 1.25 MeV [16].

Using the data from Table 1 and equation (11), the mass attenuation coefficient for the test object material can be calculated by

$$\mu_\rho = \omega(H)\cdot\mu_\rho(H) + \omega(O)\cdot\mu_\rho(O) + \omega(C)\cdot\mu_\rho(C) + \omega(N)\cdot\mu_\rho(N) + \omega(Ba)\cdot\mu_\rho(Ba) = 0.056 \text{ cm}^2/\text{g}.$$

The linear attenuation coefficient of the test object material is calculated by equation (10)

$$\mu = \mu_\rho \cdot \rho = 0.056 \cdot 1.8 = 0.1013 \text{ sm}^{-1}.$$

The initial attenuation coefficients for each hole are determined in a similar manner and are equal to: μ_0=0.3, μ_1=0.1, μ_2=0.1, μ_3=0 and μ_4=0, respectively.

The equation similar to (9) can be used to calculate the detector radiation intensity in Mathcad

$$P = P_0 \cdot e^k, \quad (11)$$

where $k = -L1(x_i, \varphi, r0, r_i)$, (13); $x_i = -r0 + 2\frac{r0i}{n}$ are coordinates of an arbitrary point from –r0 to r0 cm [19, 24].

Table 1. Content of chemical elements and their mass attenuation coefficients for $Ba(NO_3)_2$+ $C_6H_{10}O_5$

Element	Percentage, ω, %	Mass attenuation coefficient, μ_ρ, cm^2/g
H	3	0.114
O	41	0.054
C	15	0.062
N	7	0.055
Ba	34	0.052

The scanning angle range φ from 0 to 2π is used to perform calculations. Figures 2–5 illustrate changes in the detector radiation intensity when the point source changes its position within a range from –30 to 30 cm (the test object size in the section plane) for scanning angles 0, π/4, π/2 and π, respectively.

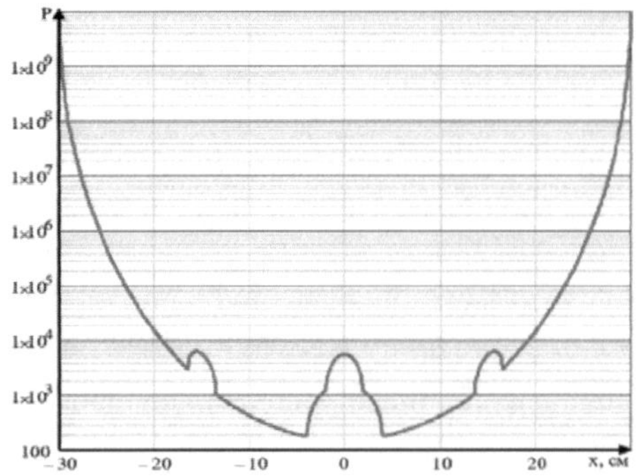

Figure 2. Graph of the change in the γ-quantum flux intensity when the scanning line changes its position from –30 to 30 cm for scanning angle $\varphi = 0$

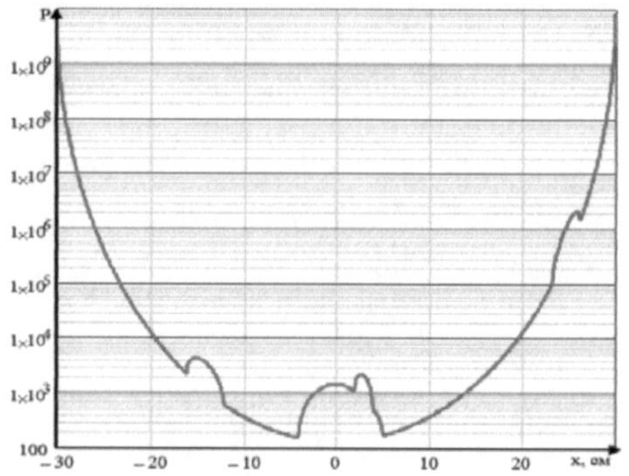

Figure 3. Graph of the change in the γ-quantum flux intensity when the scanning line changes its position from –30 to 30 cm for scanning angle $\varphi = \pi / 4$

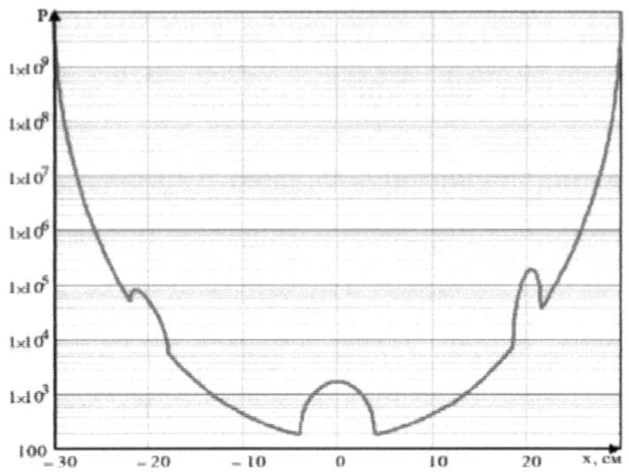

Figure 4. Graph of the change in the γ-quantum flux intensity when the scanning line changes its position from –30 to 30 cm for scanning angle $\varphi = \pi / 2$

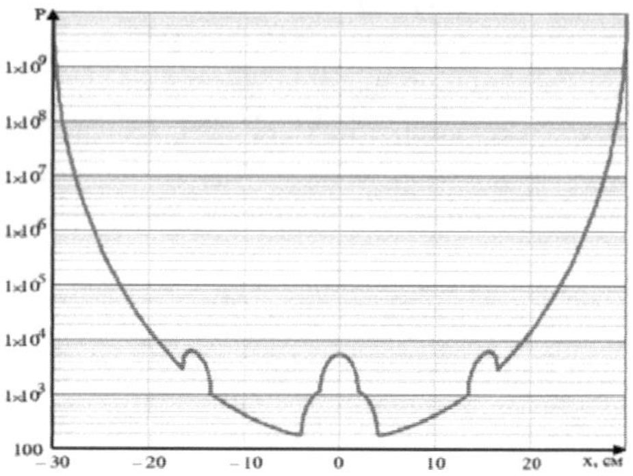

Figure 5. Graph of the change in the γ-quantum flux intensity when the scanning line changes its position from –30 to 30 cm for scanning angle $\varphi = \pi$

The graphs show that the presented calculations can be used to easily detect the position of internal holes and to determine their size and structure (by analyzing the attenuation coefficient). The graphs for different scanning angles make it possible to reveal the internal structure of the object scanned.

5. Conclusion

1. The mathematical relationships describing the length of the absorption line in cylindrical test objects are defined. Test objects containing an arbitrary number of standard irregularities of circular shape are used for scanning.
2. The obtained mathematical relationships describing the length of the absorption line of γ-quantum coming from Co^{60} isotope are corrected, taking into account the attenuation coefficient for the material with known chemical composition and density.
3. An algorithm for calculating the length of radiation ray passing through the test object with an arbitrary number of standard irregularities of circular shape, which is made of the material of the known chemical composition and density, is provided.
4. Test calculations are performed for several projection angles to show the possibility of modeling the process of collecting projection data for arbitrary sizes and locations of inhomogeneity's in the test object.

References

[1] Hussein E M A 2011 *Computed Radiation Imaging. Physics and Mathematics of Forward and Inverse Problems* (Elsevier: Netherlands)
[2] Gradoboev A V *and* Orlova K N 2014 *Adv Mater Res* **880** 237–241 doi: 10.4028/www.scientific.net/AMR.880.237
[3] Payuk L A *et al* 2016 *J Phys: Conf Series* **671(1)** 012044 doi: 10.1088/1742-6596/671/1/012044
[4] Karlova G F *and* Gradoboev A V 2013 *Proc of SIBCON'2013* 6693581 doi: 10.1109/SIBCON.2013.6693581
[5] Surzhikov A P *et al* 2016 *IOP Conf Ser: Mater Sci Eng* **110(1)** 012002 doi: 10.1088/1757-899X/110/1/012002
[6] Plotnikova I V *et al* 2018 *IOP Conf Ser: Mater Sci Eng* **289 (1)** 012029 doi:10.1088/1757-899X/289/1/012029
[7] Hemmy D et al 1983 *Neurosurgery* **13(5)** 534–541
[8] Vavilova G V *and* Ryumkin A V 2018 *IOP Conf Ser: Mater Sci Eng* **289 (1)** 012017 doi: 10.1088/1757-899X/289/1/012017
[9] Natterer F 1990 *The mathematics of computerized tomography* (Stuttgart: John Wiley & Sons)
[10] Grangeat P 1990 *Proc of Conf Mathematical Methods in Tomography* (Oberwolfach, Germany) 66–97
[11] Carmignato S *et al* (eds) 2018 *Industrial X-Ray Computed Tomography* (Springer International Publishing AG: Switzerland) 185–228 doi: 10.1007/978-3-319-59573-3_5
[12] Brenner D J *and* Hall E J 2007 *N Engl J Med* **357** 2277–2284
[13] Nizhegorodov A I *et al* 2017 *IOP Conf Ser: Mater Sci Eng* **289(1)** 012014 doi:10.1088/1757-899X/289/1/012014
[14] Coles M E *et al* 1994 *Proc of Annual SCA Meeting* 12–14
[15] Plotnikova I *et al* 2018 *MATEC Web of Conferences* **155** 01052 doi: 10.1051/matecconf/201815501052
[16] Storm E *and* Israel H 1967 *Photon cross sections from 0,001 to 100 Mev for elements 1 through 100* (Los Alamos: Los Alamos Scientific Laboratories)
[17] Vishnyakov G N *et al* 2004 *Microscopy and Analysis* **18 (1)** 15–17
[18] Hansell D M 2001 *Eur Radiol* **9** 1666–1680
[19] Bakulin V N *et al* 2015 *J Eng Phys Thermophys* **88(2)** 556–560 doi: 10.1007/s10891-015-1221-7
[20] Sarsikeev Y *et al* 2014 *Advanced Materials Research* **953–954** 529–532 doi: 10.4028/www.scientific.net/AMR.953-954.529
[21] Uglov V V *et al* 2013 *Russ J Non Ferr Met+* **54(4)** 349–354 doi: 10.3103/S1067821213040159
[22] Gavrilin A *et al* 2016 *MATEC Web of Conf* **79** 01078 doi: 10.1051/matecconf/20167901078
[23] Yan X *et al* Leahy R 1992 *Phys Med Biol* **37 (3)** 493–506
[24] Shishlov A V *et al* 2000 *Physics of Plasmas* **7(4)** 1252–1262 doi: 10.1063/1.873936

The architecture of adaptive neural network based on a fuzzy inference system for implementing intelligent control in photovoltaic systems

R Gimazov, S Shidlovskiy
Tomsk State University, Tomsk, Russia
E-mail: ruslgim@gmail.com

Abstract. In this paper, we consider the architecture of the algorithm for extreme regulation in the photovoltaic system. An algorithm based on an adaptive neural network with fuzzy inference is proposed. The implementation of such an algorithm not only allows solving a number of problems in existing algorithms for extreme power regulation of photovoltaic systems, but also creates a reserve for the creation of a universal control system for a photovoltaic system.

1. Introduction
Photovoltaic systems have a relatively low efficiency. One way to increase their efficiency is to implement extreme power control. The MPPT (maximum power point tracking) algorithms allow minimizing the underproduction of power. It is known that the existing algorithms of extreme control have a number of disadvantages related to power fluctuations, losses when entering the work point, dependence on system configuration or external conditions. Eliminate these problems by adapting intelligent algorithms [1], for example algorithms based on fuzzy logic [2]. In this paper, we consider the architecture of an algorithm for extreme regulation in a photovoltaic system based on an adaptive neural network with fuzzy inference. The implementation of such an algorithm not only allows solving a number of these problems, but also creates a reserve for the creation of a universal control system for a photovoltaic system.

The urgency of solving these problems is related to a number of tasks, including the task of maximizing the time for the operation of autonomous objects, the task of minimizing battery charge time, the task of increasing the reliability of autonomous power supply systems, and the task of creating a universal system.

The purpose of this work is the development of an algorithm for extreme regulation in the photovoltaic system based on an adaptive neural network with fuzzy inference.

The tasks solved in the course of the development work include the design of the architecture of the adaptive neural network with fuzzy inference and its modeling.

2. The ANFIS architecture for the extreme regulation of photovoltaic systems
Adaptive neuro-fuzzy inference system, ANFIS is an artificial neural network based on the fuzzy Takagi-Sugeno output system. This method integrates the principles of neural networks with the principles of fuzzy logic, which allows to combine their advantages in one structure.

The introduction of ANFIS as a control device will solve the problems associated with the nonlinearity of functions describing processes in the photovoltaic system and eliminate the drawbacks of conventional algorithms for extreme control.

Figure 1 shows the structure of a photovoltaic system with an ANFIS controller.

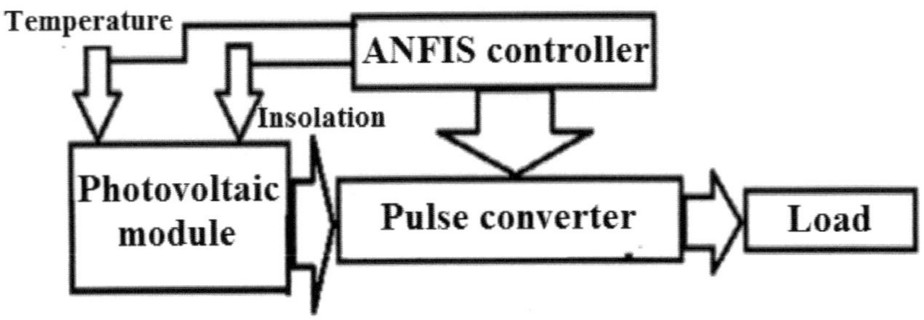

Figure 1. Structural diagram of the photovoltaic system with ANFIS controller.

In the presented system, extreme power regulation is carried out by means of a pulse converter: when the voltage / current of the photoelectric module increases, the ANFIS controller increases the pulse frequency, which in turn leads to an increase in the output current. By increasing or decreasing the current / voltage in the system, the controller goes to the optimal pair providing the maximum power.

Figure 2 shows the algorithm for applying ANFIS in the system.

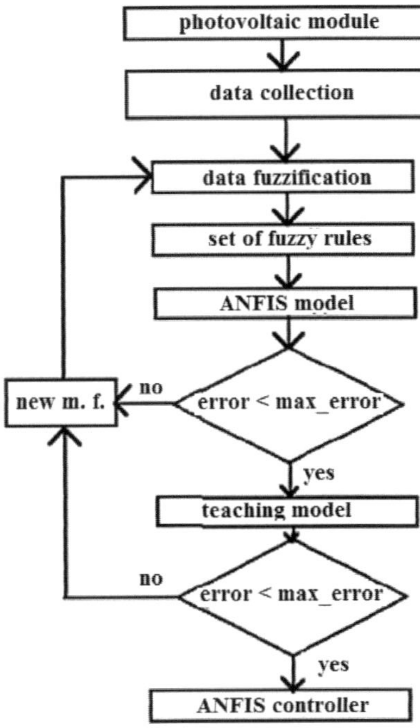

Figure 2. ANFIS algorithm.

As input variables, ANFIS uses the temperature and insolation of solar cells, because these two factors have the greatest impact on energy production. These variables undergo fuzzification procedure by membership functions. Then follows the processing of data on a set of fuzzy rules [3]. The output parameters are compared with the pre-loaded reference model. In the case of data mismatch, the membership functions are reset. In the case of data reconciliation, the model becomes a learning model.

The structure of ANFIS is shown in Figure 3.

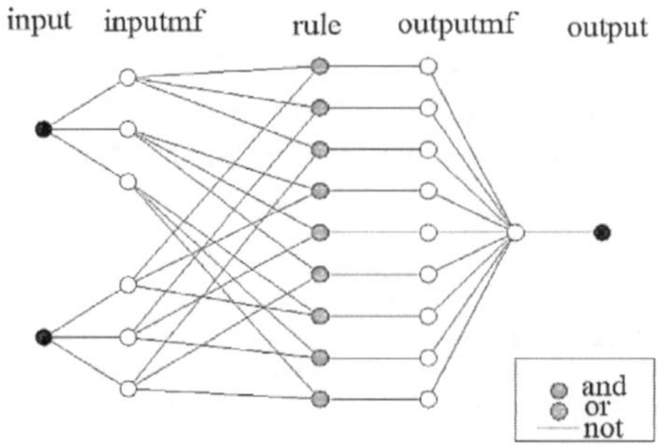

Figure 3. Structure of ANFIS.

ANFIS implements the Sugeno fuzzy inference system in the form of a five-layer neural network of direct signal transmission. The purpose of the layers is as follows:

the first layer is the terms of the input variables;

the second layer - antecedents (parcels) of fuzzy rules;

the third layer - the normalization of the degree of implementation of the rules;

the fourth layer - the conclusion of the rules;

the fifth layer is the aggregation of the result obtained by different rules.

For the linguistic evaluation of input variables (temperature and insolation), three terms are used, connected by nine fuzzy rules. The output signal is formed taking into account the execution of all fuzzy rules.

3. Simulation model of the system

To create a simulation model, the MATLAB / Simulink environment was used. To assess the efficiency of ANFIS, a photovoltaic system has been developed. The model includes a solar battery, a charge controller, a battery and a load. The developed model is shown in Figure 4.

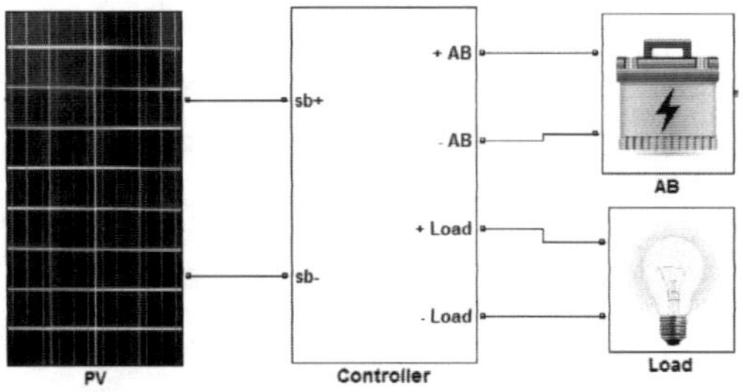

Figure 4. Model of the photovoltaic installation.

The simulation model of the solar battery is realized on the basis of the model of a single diode, with expansion to a specified number of solar cells in a solar battery. This model takes into account various factors of influence (temperature, illumination, angle of incidence of rays) [3].

Figure 5 shows the implementation of the model in the Simulink environment.

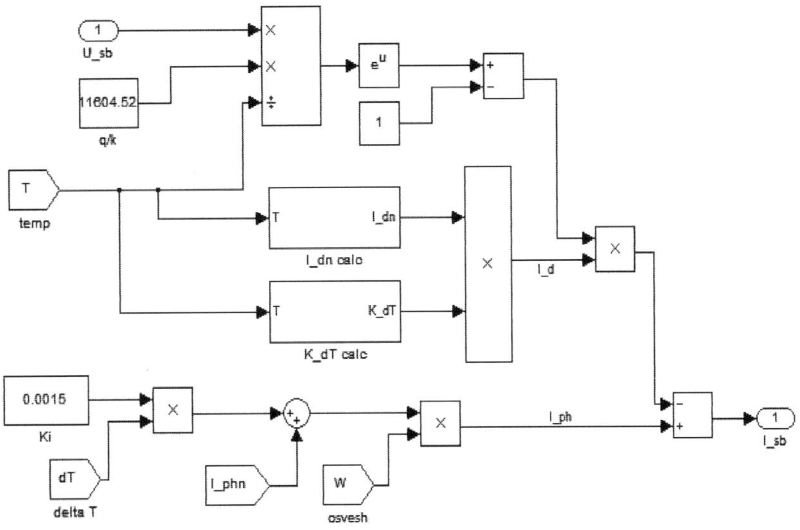

Figure 5. Solar cell model.

The controller model includes a pulse converter and a control device with ANFIS. The implementation of the model in the Simulink environment is shown in Figure 6.

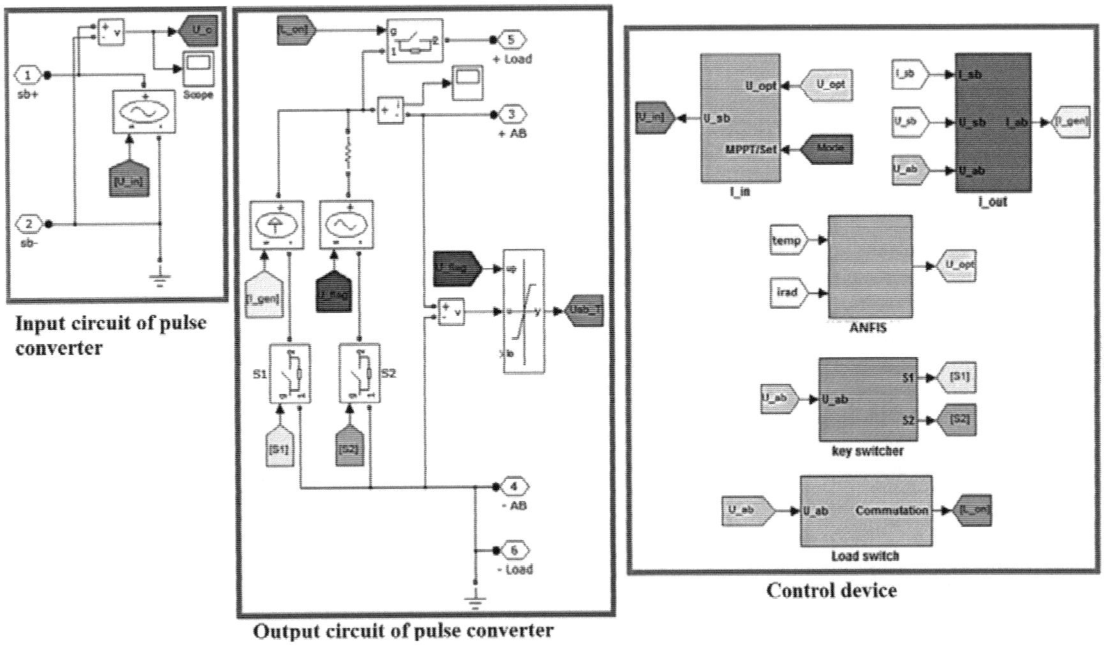

Figure 6. Model of the control system.

4. Results/ Investigation of the influence of ANFIS on the energy efficiency of the system

To create a simulation model, the MATLAB / Simulink environment was used.

To study the effect of ANFIS on the energy efficiency of the photovoltaic system, a comparative analysis of the performance of a conventional extreme algorithm, an algorithm of extreme regulation with fuzzy logic [4], and an ANFIS algorithm was carried out. The research was carried out on the basis of the developed simulation models. The results are shown in Figure 7.

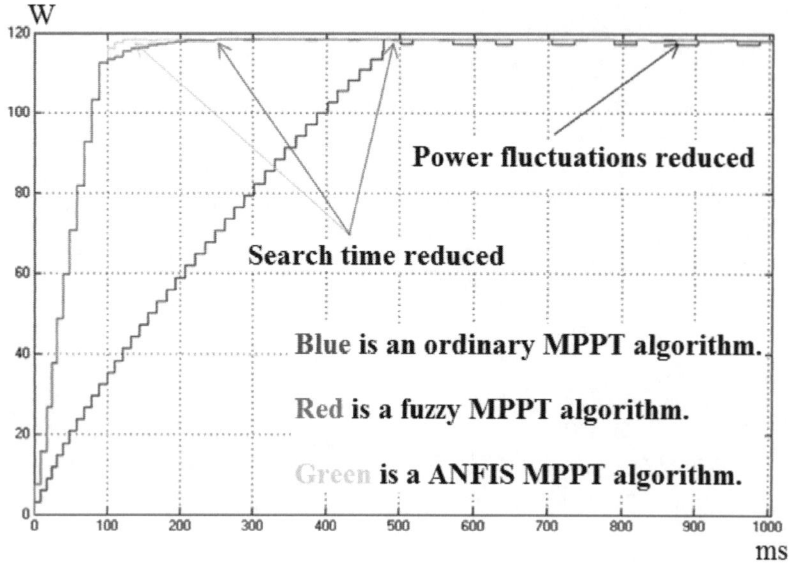

Figure 7. Search for the maximum power point by different algorithms.

Figure 7 shows the main advantages of using the ANFIS algorithm: reducing the search time of the operating point, neutralizing the power fluctuations.

5. Summary

The use of ANFIS significantly improves the characteristics of the MPPT algorithm: decreases the search time for the extreme point, decreases the amplitude of voltage fluctuations, reduces the underproduction of power.

- Reduces the search time of the extreme point by 42%.
- Power fluctuations decrease or are neutralized (depending on the number of epochs).
- The calculated efficiency values show an increase in the efficiency of the photovoltaic system by an additional 4–5% when using the MPPT controller with ANFIS.

Acknowledgments

The research was carried out in Tomsk State University with the financial support of the Ministry of Education and Science of Russia, a unique project identifier: RFMEFI57817X0241.

References

[1] Yurchenko A, Syriamkin V, Okhorzina A, Kurkan N 2015 *IOP Conference Series Materials Science and Engineering* **81(1)** 012097

[2] Shidlovskiy S 2016 *MATEC Web of Conferences* **79** 01014

[3] Panin S V, Shakirov I V, Syryamkin V I, Svetlakov A A 2003 *Avtometriya* (1) 37–53

[4] Shidlovskiy S, Gimazov R 2016 *Youth and Modern Information Technology* **1** 157–158

IOP Conf. Series: Materials Science and Engineering **363** (2018) 012017 doi:10.1088/1757-899X/363/1/012017

Development of software-hardware complex for investigation of the vector field of speeds in the cyclone-separator

A Borisov
Altai State Technical University named by I.I. Polzunova, Barnaul, Russia
E-mail: boralp@mail.ru

Abstract. The current issue of studying the vector velocity field in a cyclone-separator with a screw insert is considered in the article. Modeling of the velocity vector field in SolidWorks was carried out, tangential, axial and radial velocities were investigated. Also, a software and hardware complex was developed that makes it possible to obtain data on the speed inside a cyclone separator. The results of the experiment showed that on flour dusts the efficiency of the cyclone separator in question was more than 99.5%, with an air flow rate of 376 m^3 / h, 472 m^3 / h and 516 m^3 / h, and ΔP less than 600 Pa. The velocity in the inlet branch of the screw insert was 18-20 m / s, and at the exit of the screw insert the airflow velocity is 50-70 m / s.

1 Introduction

Automation of technological processes is the basis in modern flour milling. Increasing labor productivity, ensuring the safety of production and personnel, increasing the yield of products is possible only with the use of an automatic control system. With the use of modern automation equipment, it is possible to extend the service life of the equipment by 10-15 years [1].

Federal laws "On Environmental Protection" of 10.01.2002 No. 7-FZ; No. 96-FZ of 04.05.1999 "On the Protection of Atmospheric Air" regulate the necessity and procedure for the use of gas purification facilities at sites that have a negative impact on the environment.

At present, cyclones are the basis for dedusting flour mills. For dry cleaning of large volumes of air inertia-gravitational dust separators (cyclones) are used. Their structural elements provide rotational or translational motion of the air flow.

In comparison with other types of dust separators, cyclones have the following advantages: simplicity of design, reliability; satisfactory efficiency, durability and maintainability; high throughput, as well as relatively low aerodynamic drags [2].

It should be noted that the real efficiency of air purification in cyclones under industrial conditions is much lower (of the order of 80%), which is due to various reasons; for example, it may be not fulfilling the condition of the matching of the input speed to the optimum value [3, 4].

The development of a cyclone separator on the basis of an air-screw insert located on the cyclone exhaust pipe and located in a confuser with a perforated surface made it possible to create a fundamentally new method [3], which has no analogues. In the laboratory, currently has an efficiency of 95-99.9%, which is higher than foreign analogs.

The aim of the work is the development of a software and hardware complex for the investigation of the vector velocity field in a cyclone-separator.

The tasks of the work include:

- to build and conduct research on the mathematical model in SolidWorks;
- develop a software and hardware system for the investigation of the vector velocity field;
- to carry out research on the experimental installation of the cyclone-separator with the help of the developed complex;
- compare the results obtained during the study.
- control over the movement of personnel and clients of the organization both inside and outside the building;
- obtaining data for information and analytical work.

2 Principle of operation

To increase the efficiency of the cyclone, it is necessary to control its technical parameters, such as the degree of purification, the volume of air to be purified, and the pressure loss.

In most cases analogue, as well as portable digital devices are used for these purposes.

The disadvantages of analog devices include the accuracy, which affects the wear of moving parts and the human factor. Portable digital devices of these drawbacks are lacking, but the high price, the lack of simultaneous monitoring of the set of technical parameters, and in some cases the duration of measurements, make these devices unacceptable for continuous monitoring of parameters [4].

The functional diagram of the experimental dust separator is shown in Figure 1 [5, 6].

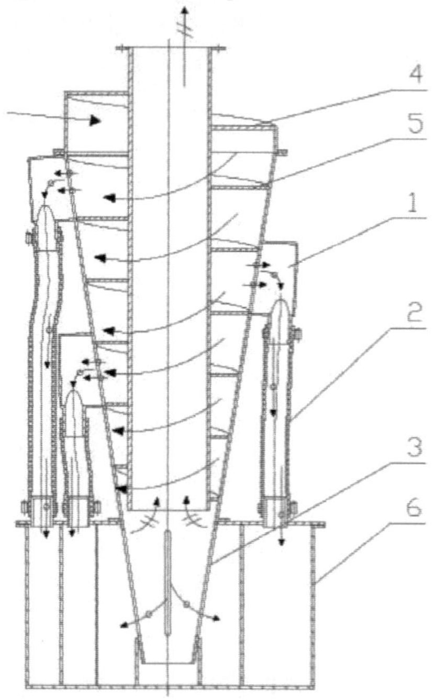

→ - motion of the raw air, → - motion of clean air, → - movement of dust.
1 - the box; 2 - flexible hose; 3 - attachment; 4 - snail; 5 - screw insert; 6 - container
Figure 1. Functional diagram of the experimental cyclone-separator.

The air mixture on the material conduit enters the input cochlear 4 of the experimental dust separator. The input snail serves to communicate the rotational axisymmetric motion to the product and thereby create conditions for the separation of the product from the stream. Exiting the swirler, the flow enters the conical part of the experimental dust separator and acquires additional rotary screw motion due to the screw insert 5.

Large particles of the product rush to the walls of the experimental dust separator due to the effect of the centrifugal force that occurs in the rotating stream, as a result of which their main mass enters the slot to catch large dust and precipitates in boxes located at different heights. The rest of the dust, along with the air, moves axially to the lower part of the dust separator, where, under the action of inertia, the dust gets into the cone-cone 3 for additional cleaning of the air, and the purified air leaves the exhaust pipe at the lowest resistance.

Changing the gap between the cochlea to which the screw insert is fixed, and the resistance of the dust separator changes with the outer cone. Within the acceptable limits, lifting the cochlea does not significantly change the efficiency of the dust removal.

To improve sealing, sealing gums are used, sealant on all mating parts of the dust separator.

3 Results of device testing

To study the vector velocity field in a cyclone-separator, a model was developed in SolidWorks [7] (Figure 2).

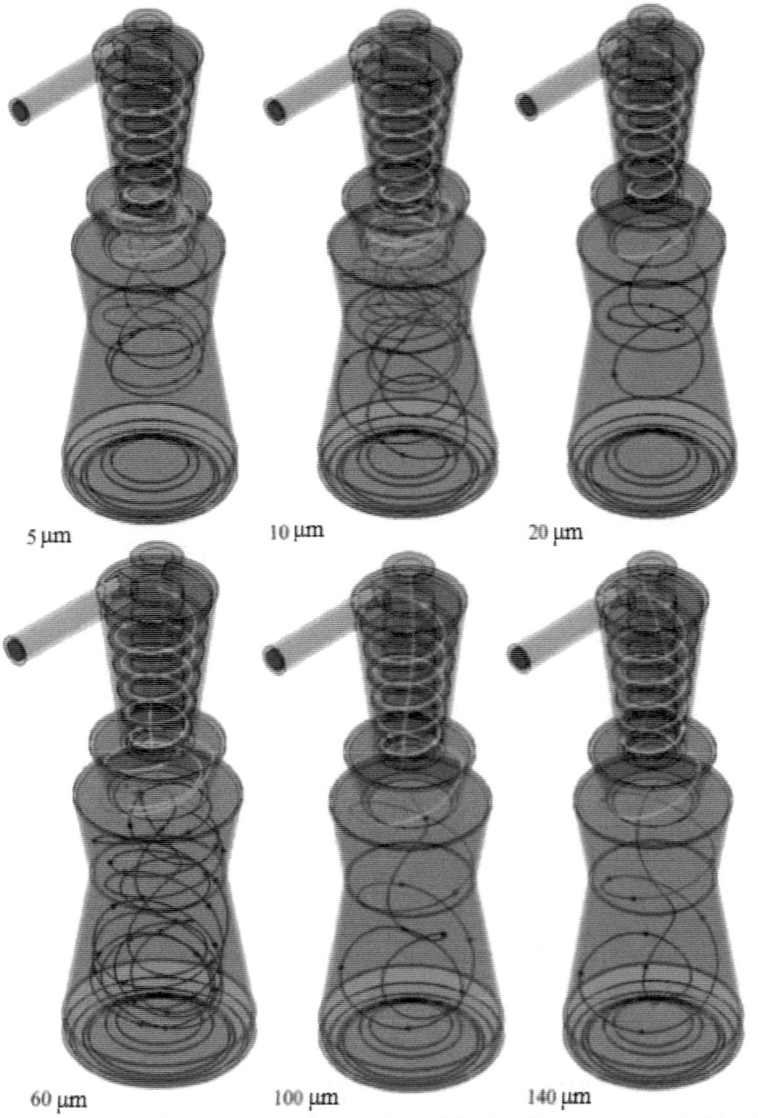

Figure 2. Dependence of the motion of particles of finely dispersed material in the apparatus.

It can be seen from Figure 2 that particles, due to a large radial velocity, push against the wall of the cone.

Figure 3 shows the change in radial velocity.

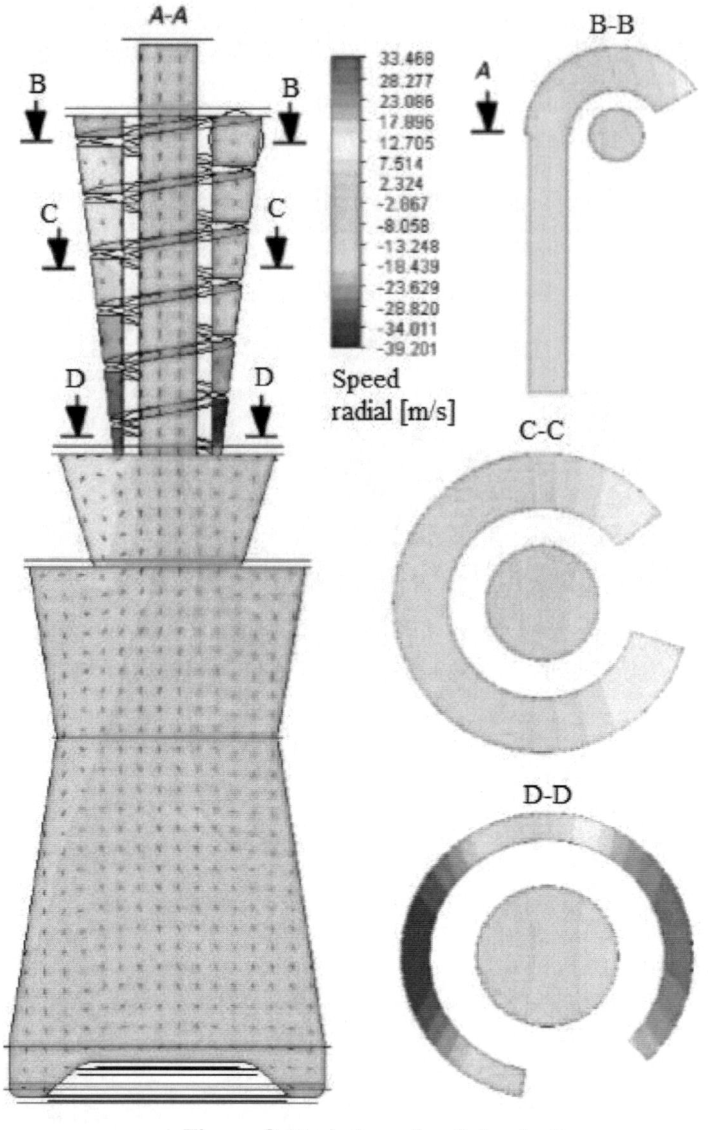

Figure 3. Variation of radial velocity.

Figure 3 shows that the velocity distribution is also not symmetrical and curved. This characterizes that when the air moves through the screw insert, the velocity increases, and the flow is concentrated on the outer wall of the cone. There is a coincidence with the field of dynamic pressure, which corresponds to the fact that this type of speed is predominant in this installation.

The software-hardware complex consists of (Figure 4) a cyclone separator, a Raspberry Pi microcomputer and a client application [8].

A complete model of the information flows of the system for measuring the velocity vector field is shown in Figure 5.

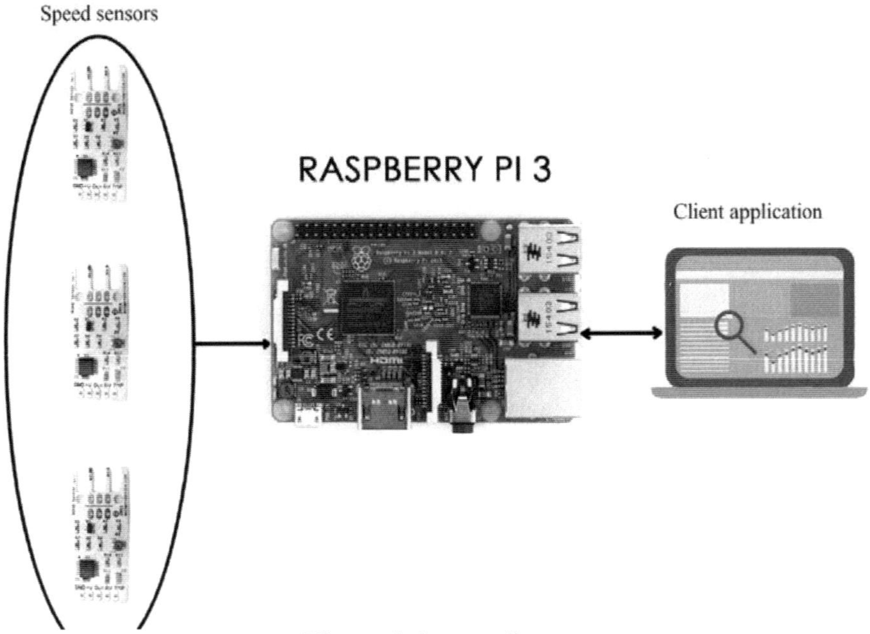

Figure 4. System diagram.

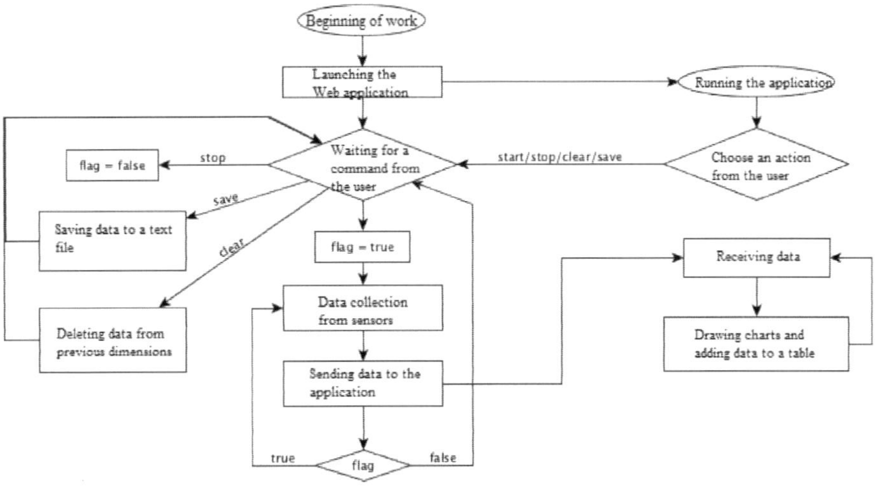

Figure 5. Information flows.

The software-hardware complex works as follows [9, 10]:

- after launching the application by the user, provided that the whole system is working properly, the server application and the client part will be launched, which will be available in the localhost: 3000;
- further the user selects the action that he wants to perform - start the system, stop, save the data in a text file for further processing or clear all data. At the first moment of the start it makes sense to choose the launch, which the user is allowed and selected at the moment;
- the command comes to the server, the server starts to collect data from five wind sensors and store them for further storage and send it to the client application for graphical representation;
- the client application receives data, first processes it, and then adds a new row to the table and new data to the graphs;

- in this mode, the application will work until the user stops the system by pressing the stop button. When he does this, a request for a stop will be sent to the server. All data is saved, only the receipt of new data stops;

- then, the user has a choice, either wipe the data, or save them. If the data is not overwritten, and run another experiment, the new data will be added to the old one.

Figure 6 shows the graphs of the dependence of the speed obtained from the wind sensors on time.

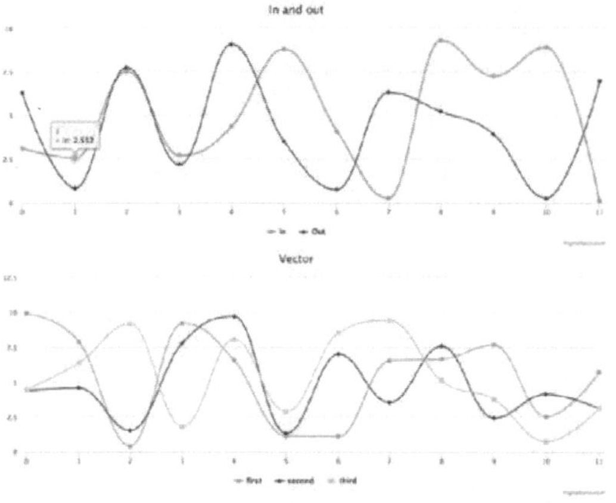

Figure 6. Graphs of the velocity vector field.

Conducted laboratory tests using the software-hardware complex for a cyclone separator showed very satisfactory results both in terms of efficiency and in the aerodynamic resistance of the cyclone. On flour dusts, the efficiency of the cyclone separator in question exceeded 99.5%, with an air flow rate of 376 m^3 / h, 472 m^3 / h and 516 m^3 / h, and ΔP less than 600 Pa. The velocity in the inlet branch of the screw insert was 18–20 m / s, and at the exit of the screw insert the airflow velocity is 50–70 m / s.

4 Conclusion

When comparing the model in SolidWorks and the data obtained with the software-hardware complex, it can be concluded that the discrepancy in the results is not more than 1%. Using modern means of automation, it is possible for 10 - 15 years to extend the life of technological equipment. But the main thing is that without modern automatic control systems, it is impossible to guarantee the quality of the products.

References

[1] Gupta A, Lilly D, Sayred N 1987 *Swirling flows* (Moscow, Mir)

[2] Straus V 1981 *Industrial purification of gases* (Moscow, Chemistry)

[3] Saputro H *et al* 2018 *IOP Conf. Ser.: Mater. Sci. Eng.* **288** 012142 doi:10.1088/1757-899X/288/1/012142

[4] Sastry S 2013 *Performance Studies of Cyclone Separator: Experiment and Design Considerations Paperback use pre formatted date that complies with legal requirement from media matrix by Susarla Venkata Ananta Rama Sastry S. Sastry* (LAMBERT)

[5] Zlochevskiy V L Aerovintovoy tsiklon-separator [Aerovoltaic cyclone separator], Patent RF № 2442662, 2012 (Russian)

[6] Xu W W, Li Q, Zhao Y L, Wang J J and Jin Y H 2016 *IOP Conference Series: Materials Science and Engineering* **129** 012022

[7] Borisov A P 2017 *Technologies of food and processing industry of AIC - healthy food products* **4** 88

[8] Zlochevsky V L, Mukhopad K A 2015 *South-Siberian Scientific Bulletin* **4** 5

[9] Shvab A V, Zyatikov P N, Sadretdinov Sh R, Chepel A G 2010 *Applied Mechanics and Technical Physics* **51** 174
[10] Nikolaeva V K, Borisov A P Program for controlling and controlling airflow in the cyclone separator Certificate of state registration of the computer program №2017662653, filed. 26.09.17, publ. 11.13.17

IOP Conf. Series: Materials Science and Engineering **363** (2018) 012018 doi:10.1088/1757-899X/363/1/012018

Reflex inhibition. Global manifestation of the local mechanism

V Shumilov, V Syryamkin
National Research Tomsk State University, Tomsk, Russia
Email: vnshumilov@rambler.ru

Abstract. The paper considers the issues of training and retraining in self-organizing natural information processing systems and shows the reasons for the rule "The one who comes first gets all" and the possibility of creating a technical device, which is an artificial intelligence carrier. Such a device is not filled with the necessary information at the stage of creation, but receives all the necessary information in the process of functioning, especially at the beginning of its activity and in the process of education, just as the intellect of a person is being raised since childhood.

1 Introduction

Until now, physicians and biologists speak of "the doctrine of the higher nervous activity" when describing the functioning of the nervous system (NS) and the brain [1–4] (I.Setchenov, I.Pavlov). At the same time, specialists in other fields speak about their disciplines as theories. It is generally accepted that a theory differs from teaching (doctrine) by its logical validity and practical confirmation. Whereas a doctrine is based mainly on believe in the truth of a doctrine based on the authority of a doctrine's creator without special concern about practical confirmation of a doctrine and based on the absence of alternative evidentiary logical constructions that resolutely reject the doctrine. Why does this situation continue to be true? When applied to the discipline concerning such an important subject as conditioned reflexes and, ultimately, our thinking, the use of the term "doctrine" indicates the absence of a firm theoretical and practical foundation for this discipline despite achievements in the field of cognition and successes in correction of the nervous activity when necessary. Indeed, there are still no theoretically substantiated and practically confirmed representations about the mechanisms of functioning of memory, thinking, and about the details of these mechanisms and processes. The aims of the work presented by us include change of the situation in the study of higher nervous activity; we mean the change of the term "doctrine" to the term "theory". We will try to do this based on the principles of brain functioning, which are theoretically grounded [5], supported, and patented [6-9]. The principles proposed were confirmed practically during implementation of the research project RW 9006 [9], but so far only by the example of a small electronic scheme. The model of the simplest nervous system consisting of three electronic neural-like elements demonstrated the ability to completely independent "learning" by self-formation of connections between neurons, in which traces of events are reflected. In addition, this simple model demonstrated the ability to re-learn when circumstances change. It is expected to confirm the principles proclaimed by conducting experiments on simple organisms (by the example of nervous systems of simple biological objects such as mollusks and insects).

2 The mechanism of training (formation of communication between neurons)

The main features of brain functioning described in [5] include a purely local mechanism for formatting connections between pairs of excited neurons and a continuous decrease in the weights of all connections (new and long-existing). Only the state of neurons and their spatial proximity (mechanical contact) have the meaning for formation of connections between neurons. The mechanism of formation of connections should not only be purely local but also extremely simple because it originated itself, it was not designed by engineers. Confidence in the simplicity of the mechanism for formation of inter-neural connections follows already from the fact that when the organism is born, there are no connections between neurons, no mechanism for establishing connections, and even no neurons themselves. There is no place for storage of information about the mechanism for control over formation of links and about rules, according to which communication should be established. Where is the information about this local mechanism arrangement stored? This information is contained in the parent (mother) cell itself; it includes the information about how a cell can reproduce its copy next to itself, how it reacts to the nearest environment, which, in turn, can influence it. This mechanism ensures formation of connections between mechanically contiguous components of neuron pairs throughout the human brain with an average speed of about 30 thousand new connections per second. The rate of formation of connections is easily estimated by the time of life of the brain and the number of connections between neurons in the mature brain. The number of 30 thousand comes from the fact that the number of connections in the human brain during its lifetime (about 100 years) grows from zero at conception to the value of about 150×10^{12} connections (15 billion neurons, up to 10 000 connections for each neuron). The proposed mechanism for local formation of connections is realization of the logic of formation of connections between excited neurons at the intersection of half-selected sets of outputs of *exciters* and inputs of *excited neurons*. This can be represented visually as appearance of a standard value positive potential at the outputs and a negative potential of approximately the same value at the inputs of the excited neurons and further appearance of the possibility of forming connections between points if the potential difference of these points exceeds the potential for forming connection U (FP). The scheme of formation of the connection is shown in Figure 1.

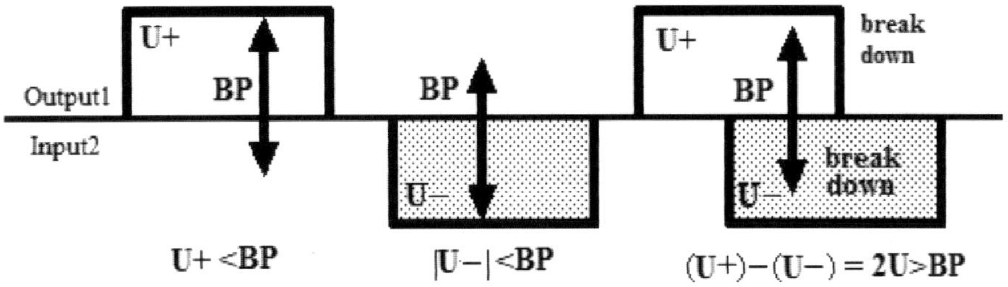

Figure 1. Formation of the connection between two excited neurons.

This Figure illustrates the logic for formation of a connection between excited active elements (neurons in a biological organism or neuron-like elements (NE) in artificial devices). Excitation of active elements occurs in a usual way. It occurs when the total signal at the input of the active element exceeds a certain threshold Uth.

Formation of the depicted potentials at the outputs and inputs of the excited elements can be easily realized for electronic devices. Such formation of potentials contradicts the usual practice of creating various electronic devices. However, it provides formation of connections between the NE of the electronic device. Is existence of such mechanism possible in natural organisms?

Let us consider appearance of an electrical potential at the output of a biological neuron. Neurons are electrically isolated from the environment well enough. When excited, the neuron produces a short pulse, and then remains in an unexcited state for a while even if a powerful and long-lasting input signal effects it at this time. On the one hand, this leads to generation of a series of short pulses at the output of the neuron. On the other hand, this indicates that the neuron can produce a powerful short-term signal because of the discharge of the charge accumulator (battery, capacitor) and the subsequent relatively slow charge of this accumulator. Loss of the neuron's efficiency for the time of charging and a series at the output when feeding a continuous signal at the input is explained by the need to charge the accumulator. So the process of forming a neuron output signal can be described as follows. The appearance of the potential at the output of the excited neuron (positive for definiteness) is provided by "closing a contact" of the internal cell accumulator battery, that is a source of the EMM (electromotive medium). Since the body of the neuron is well isolated including isolation from the earth (0 volts), the delivery of a potential to one neuron point (output) automatically leads to the displacement of the potential of the other neuron point (input), to which the potential from the opposite source clamp is given (charge storage). Therefore, if the resistance of equivalent circuits, is approximately equal on the one hand, between the output and "ground" and on the other hand, between the input and "ground", then the values of the potentials at the output and input will be approximately identical but with different signs. Therefore, the simplest EMM generator in the excited neuron will automatically generate potentials of different signs to the output and input of the neuron. This will create the necessary conditions for formation of connections between the output and the input of two excited neighbor neurons. Whereas, there will be no conditions for formation of connections for other points of different neurons, even if they have mechanical contact; this ensures formation of traces of events (links between excited neurons).

Formation of connections between arbitrary excited elements, both between artificial NEs and between biological neurons, is easily realized both in technical devices and in biological nervous systems (NS) according to the scheme presented in Figure 2.

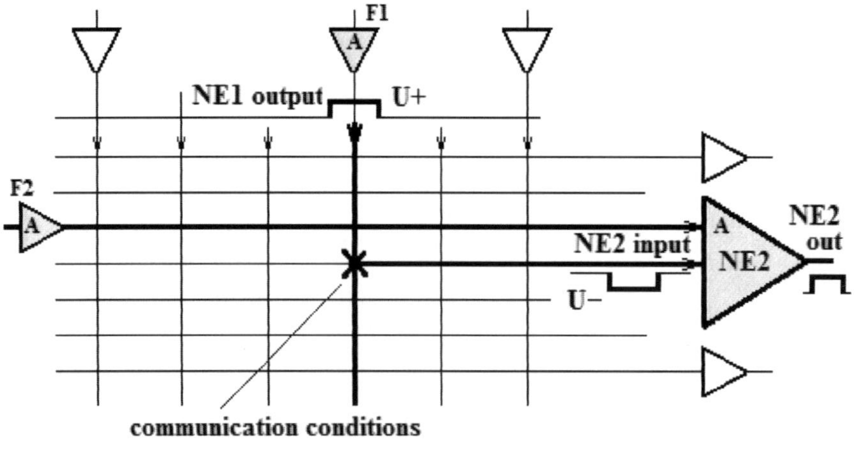

Figure 2. Formation of connections between elements of excited neurons at the inputs and outputs intersection.

Figure 2 shows the scheme for formation of connections between the neuron NE1 output and the neuron NE2 input of excited neurons, which have a potential difference sufficient to form a connection between them. Whereas the potential differences of the points of other neurons are insufficient to form connections. Here, the neuron NE1 is excited in its chain from excitation by an external factor F1 and the neuron NE2 is excited by a factor F2. If the moments of their excitation coincide, then a connection is formed completely independently (without control from the side) when a positive potential at the output NE1 and a negative potential at the input NE2 between these points appear. Therefore, subsequently, NE2 is excited at NE1 excitation through the formed connection. Because of the "forecast", NE2 will excite at NE1 excitation. Therefore, an organism can pre-evade the approaching factor2 because of the "forecast" based on the formed connection.

3 How excited neurons get tired

Neurons are not exceptional cells dealing with electricity. The receptors transform optical, chemical, mechanical, and thermal stimuli into electrical signals. Muscles turn potentials into mechanical actions. Electric generating cells (tissues) of some animals form electrical impulses of high voltage; this fact additionally indicates the general nature of all cells of the body. We should note that the cells of electric fish generating electricity rapidly weaken when generating several pulses in a row, that is, they generate a discharge of lower voltage and a lower current.

The neuron behaves the same way. Under the conditions of frequent excitement, it does not have time to recover completely and fatigue occurs. The energy for the signal is stored in the body of a neuron (in the form of nutrients). Because of excitation, the neuron each time consumes a portion of energy from its stock, which is restored at a certain rate. Therefore, the ability of the neuron to excitation decreases after each excitation. That is, the threshold voltage (UT) increases slightly. We can say that the neuron "gets tired". If a neuron is excited quite often, then its energy reserves will not have time to recover. It can even stops being excited.

4 Choice of direction of signal propagation

Because of an electrical breakdown or an equivalent phenomenon, the output of one excited neuron and the input of another excited neuron form a permanent connection for the rest of life. Communication cannot be formed with other neurons, which are not excited at the moment, because the potential differences between them are insufficient for breakdown. If the connection is formed, then, most likely, the resistance of the formed connection, as well as of the previously existing connections will slowly increase over time $\mathbf{r_{cf}} \rightarrow \mathbf{R_{cf}}$ because of the resorption of the conducting channel. Conductivity decreases, communication weakens, and the event is forgotten. Because of the weakening of ties over time, old events are recalled with greater difficulty than recent events. Recalling of old events require more associations for neuron excitation because neuron excitation occurs when the total signal, which is the sum of contributions of the signals to individual inputs, exceeds a certain threshold (UT). That is, the sum of the products of the potentials by the weights of the individual connections is greater than the threshold: $S(Ui*Wi) > UT$

We should note that when a new connection is formed, or when an element of the next stage is excited, simple rectangular potentials cease to be regular rectangular because of partial compensation of the potentials of the associated outputs and inputs. This process is shown in Figure 3.

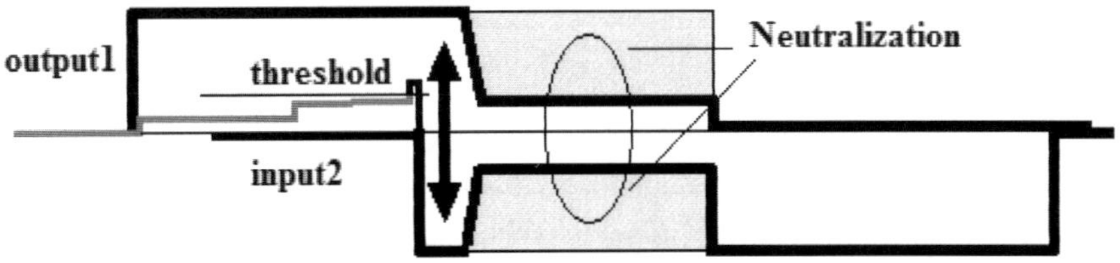

Figure 3. Compensation of potentials at the inputs and outputs of excited neurons.

Formation of bi-polar potentials at the output and input of excited neurons not only ensures formation of new connections. The opposite polar potentials mutually compensate themselves and, thus, provide a solution of other problems associated with the signal propagation through the labyrinths of brain structures. Because of the spread of the neuron parameters and weights of their input connections, the moments of exceeding the excitation thresholds will differ for different neurons and one of the neurons will be excited first. It will immediately give a negative potential to its total input. This potential partially compensates the potential that has excited this neuron. The same exciting potential, which had excited the first of the neurons, also enters inputs of other neurons. Partial compensation of the exciting potential will prevent excitation of other neurons that did not have time to be excited. Therefore, only the neuron excited first will be excited because of the optimal configuration of the weights of its input connections and the values of the input signals. Because of compensation of multi-polar signals, the rule “The one who comes first gets all” is automatically generated; in the artificial neural networks, this rule is introduced by directive.

We have considered the local mechanism for formation of a trace of an event, that is, formation of a neuronal connection that does not require external control. We can say that we have examined the procedure of “training” of the nervous system. However, in real life, the conditions of existence of the organism with the NS are constantly changing. Therefore, survival and evasion of the organism from the stimulus can be ensured only when the reaction of the organism to almost identical stimuli will change in accordance with new circumstances. It means that it is necessary that the organism can change its reaction (retrain).

5 A mechanism of retraining (inhibition of obsolete reflexes)

It turns out that the mechanism of organization of functioning of the organism’s memory (including formation of new connections (remembering), forgetting (fading of connections, reduction of their weights), and neuron tiredness because of rapid energy expenditure and slow restoration of its reserves) automatically provides “retraining” of the nervous system in conditions of circumstances change. Let us show how the reaction of the organism and the conditioned reflexes change in such cases.

When circumstances change, the old reaction of the organism no longer leads to success, to evasion from the stimulus, and to the cessation of its action. Because of unsuccessful excitations of neurons forming a chain of the old reaction, which does not lead to cessation of irritation, these neurons “get tired”, their individual thresholds of excitation grow. There appears a possibility of excitation of neurons that are not part of the chain of the old reaction. Some of them are excited, new connections and a new reaction are formed. New connections have the greatest weight, which decreases over time as well as the weight of all other connections. However, the new connections continue to have greater weight than the old ones. Therefore, subsequently, in similar situations, neurons that form a chain of a new reaction are excited first because of the greater weight of connections. This is because everything in our world is inertial. Therefore, the change in potentials at the total neuron input changes at a certain rate determined by the electrical capacitance, the weight of the individual connections (inputs), and the potentials at the outputs of the neurons of the preceding cascade participating in the excitation of the neurons of the cascade under consideration. When getting excited, neurons that form the beginning of a new reaction chain (a new reflex) partially compensate the positive potentials of the neuron outputs of the previous cascade (that excite them) by negative potentials at their inputs. It means that neurons from the old reaction chain that have not yet had time to be excited can no longer be excited. As physiologists say, the “inhibition” of the old reflex occurs.

However, according to widespread contemporary ideas, the inhibition of old reflexes that do not lead to success is ensured by the existence of a special mechanism of inhibition. The essence of this mechanism is in the fact that along with the signals involved in the excitation of neurons by summing the individual signals of the same polarity at the summing input, there are also signals of the opposite polarity, which come at the same summing inputs and prevent excitation of neurons. However, none of the researchers observed these inhibitory inputs (signals) in practice.

There must be some mechanism for suppressing the old reflex. How else can the observed suppression of the old reaction be explained? We put forward the explanation of the observed phenomenon. However, if we admit the existence of inhibitory signals, then it becomes necessary to explain a mechanism of formation of such inhibitory connections (signals). What is the logic of formation of such connections and supply of inhibitory signals to them? Who or what decides when to give such signals? Obviously, such a difficult task of determination of the necessity of inhibition requires analyzing several “global” parameters that are inaccessible to an individual neuron dependent only on the potential at its summing input. That is, the local neuron mechanism cannot ensure formation of inhibitory signals according to the adopted scheme.

The local mechanism of formation of new connections and formation of signal paths with the participation of global "inhibition" proposed by us is realized by a purely local mechanism because of the tiredness of often-excited neurons and the subsequent "competition" of neurons for a limited resource (energy of the exciting signal).

Neurons become tired because of the ineffectiveness of the old reaction; it does not lead to elimination of irritation, cessation, and consequently, excitation of the neurons of the old chain. Because of the neurons tiredness (multiple excitation), they cease to be excited and to prevent excitation of other neurons. These neighboring neurons are excited and new strong connections with the greatest weights are formed. A chain of new reaction to the environment changed is being formed.

Subsequently, the neurons that had formed the chain of a new reaction would be excited first because of the great weights of their connections. In conditions of their excitation, they will prevent excitation of the neurons of the chain of the old reflex. This is how the old reflex will be suppressed because of a purely local mechanism of comparing possible ways of signal propagation along brain structures.

6 The mechanism of thinking

Nature is economical. It does not have resources to create complex mechanisms. All objects and structures are formed by themselves in accordance with initial mechanisms and the influence of their environment. We offer a purely local, extremely simple mechanism for the neuron excitation and formation of connections, which provides training, retraining of nervous systems, and the choice of signal paths through the brain structures. This local mechanism ensures functioning not only of nervous systems of the simplest organisms, but also of the brain of higher animals, which consists of tens of billions of neurons, each of which can have up to 10 000 separate inputs (connections with topological neighbors). We have examined the most important logical principles that ensure functioning artificial devices and natural biological objects, which are capable for self-training under the influence of the environment.

Omitting consideration of less significant details of functioning of such objects, we will present a scheme of the electric brain of a developed animal shown in Figure 4. Contemporary level of microelectronics allows implementing such a scheme. In the initial state, prior to the operation of the device with such a scheme, a number of connections must be established in the device; these connections must be analogous to the connections in the brain of animals, which represents unconditioned reflexes. These unconditioned reflexes are genetically conditioned; they are formed because of the evolutionary experience of this population. Based on these genetically conditioned reflexes (initial reactions) further accumulation of individual experience occurs (formation of a set of individual conditioned reflexes of the given organism).

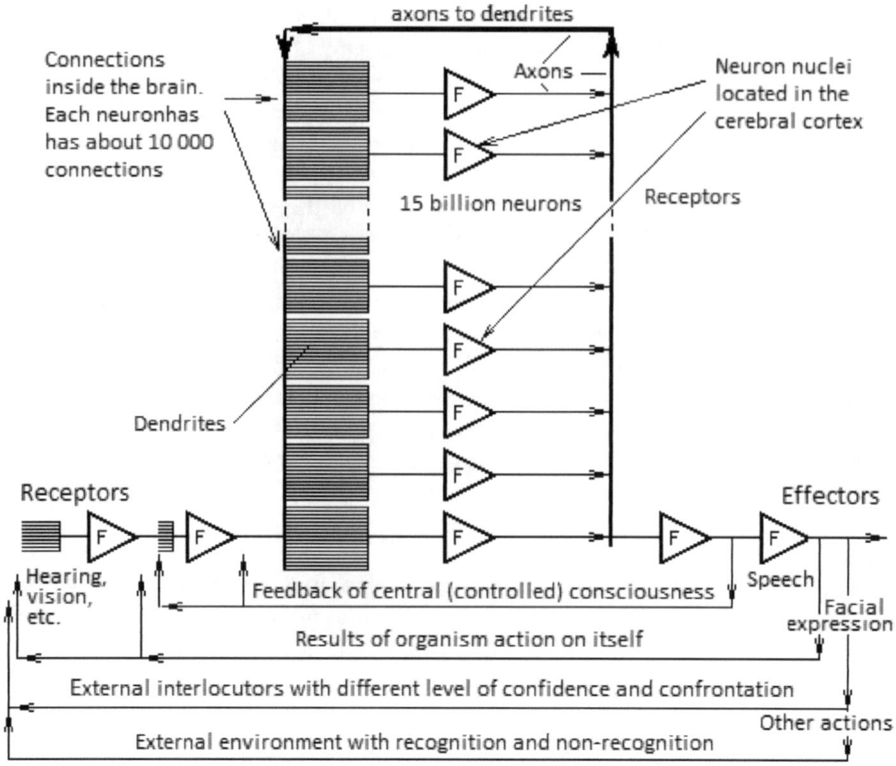

Figure 4. Electric scheme of the brain.

Accumulation of reflexes is a process of constant training and retraining with participation of a mechanism of global "inhibition". The process of constant training and retraining is formed based on the properties of the local mechanism of remembering, forgetting (formation and properties of bonds), and tiredness of neurons (active elements of retransmission of receptor responses to stimuli). Formation of conditioned reflexes in the device as well as in biological objects will occur more effectively because of training the same way as it occurs with children of various animals from the simplest to the highest ones. Therefore, in a full-scale device with the presented scheme of connections (existing and potential ones), an educated intellect can be brought up.

From the point of view of creation of devices with the presented scheme of connections between active elements - formers F, formation of reflexes is formation of chains of connections between formers (neurons, NE). Reflexes provide passage of signals from receptors to actuators (effectors) when receptors are influenced by stimuli (environmental influences) through connecting (intercalary) structures of the brain. Locality of mechanisms of brain functioning and their distribution throughout the brain volume provides parallel asynchronous modification of brain structures with a huge speed, which cannot yet be provided in sequential software and hardware complexes.

In our work, we give logical arguments. In addition to arguments about the principles of the nervous systems functioning, these principles were tested by highly qualified specialists, were protected by the patent of the Russian Federation [6], and were published and presented at various conferences [10–13]. However, most important is that these principles are confirmed in practice: we created a device - an electronic model of the simplest nervous system consisting of three neural-like elements, which has demonstrated the ability to be trained and retrained in accordance with the above-described arguments [9]. Currently, in partnership with biologists from Tomsk State University, we are trying to confirm the efficiency of the proposed principles of the NS functioning for the living biological objects.

7 Conclusion

The arguments presented in this paper demonstrate the wide possibilities of the purely local mechanism for functioning of the nervous system structures both of the simplest animals and the human brain. We can see their evolutionary connection. Realization of these principles, which formed the basis for self-organization of mechanisms of training, retraining, and thinking in natural organisms, can provide creation of artificial intellect carriers with a large number of active elements, in which an intellect equal to the human one (or even superior) can be brought up. In this situation, there is a potential danger of antagonistic competition for resources between man-made artificial devices and humans. Therefore, in order to avoid antagonism, it is necessary to create devices that will not antagonistically compete with humans, but will coexist in symbiosis and with mutual benefit.

Acknowledgments

This study (research grant No № 8.2.24.2018) was supported by the Tomsk State University competitiveness improvement programme. The authors are grateful to Tatiana B. Rumyantseva from Tomsk State University for her assistance in preparing the paper.

References

[1] Koshtoyants Kh I *Pavlov I.P. and His Theory* URL: https://www.nkj.ru/archive/articles/31937/
[2] Zhdanov A *Modeling of Higher Nervous Activity* URL: http://www.nkj.ru/archive/articles/5613/
[3] Selanin E V *Human Physiology* URL: http://human-physiology.ru/about_author/
[4] Anatomy A *Physiology and Hygiene (manual). Chapter 5. Fundamentals of the Doctrine of Higher Nervous Activity* URL: http://koi.tspu.ru/koi_books/sedokova/
[5] Shumilov V N 2008 *Principles of the Brain Functioning. Engineer's look* (Ekaterinburg)
[6] Shumilov V N, Voronin V N, Solomonov V I 2013 Patent RU (11) 2 475 843 (13) C1. Adaptive Control Device, Neural-Like Base Element and the Method of Organization of Operation of Such Device
[7] Mekhtiyev A *et al* 2016 METALURGIJA **55** 47–50
[8] Shumilov V N, Voronin V N, Solomonov V I 2012 Patent for utility model No 120800. Control Device with an Artificial Neural Network
[9] Voronin V N, Shumilov V N, Shumilov Yu A, Tsoy A R, Kiselev A V, Sharshov S M 2014 Report on research work under the contract No. 9006 dated 04.09.2014 between TSU and "Prerogative Tomsk" LLC: "Development of Electronic Circuits for Creation of an Operating Model of an Artificial Neural-Like Element", Tomsk
[10] Shumilov V N, Shumilov I V, Syryamkin V I 2015 *Neurocomputers: development, application* **(4)** 94–96
[11] Syryamkin V I , Shumilov I V, Shumilov V N 2015 *Neurocomputers: development, application* **(4)** 78–79
[12] Voronin V N, Shumilov V N, Shumilov Yu A, Tsoy A R, Kiselev A V, Sharshov S M 2015 *Neurocomputers: development, application* **(4)** 26–28
[13] Serikov M Z *et al* 2017 *Proceedings - 2016 11th International Forum on Strategic Technology, IFOST 2016* 510

Developing operation algorithms for vision subsystems in autonomous mobile robots

M V Shikhman[1,2] and S V Shidlovskiy[2]
[1]National Research Tomsk Polytechnic University, Tomsk, Russia
[2] National Research Tomsk State University, Tomsk, Russia
E-mail: Sikhmar@gmail.com

Abstract. The paper analyzes algorithms for selecting keypoints on the image for the subsequent automatic detection of people and obstacles. The algorithm is based on the histogram of oriented gradients and the support vector method. The combination of these methods allows successful selection of dynamic and static objects. The algorithm can be applied in various autonomous mobile robots.

1. Introduction

Currently, robotic systems are rightly considered as the base for automation of industry, medicine, military industry, space exploration and other spheres of human activity. Their joint work allows achieving various goals and solving a wide range of technological problems.

An autonomous mobile object is a kind of robotic system, which is based on an autonomous control system provided by optical sensors, computer algorithms and radar. An autonomous mobile object can be used to move passengers or cargo, to deliver materials, technological or other types of equipment, and to collect visual information when working with additional equipment. The main two tasks for any autonomous object are as follows. First, to be able to plan the path independently; second, to be able to move around the dynamic environment successfully, including the ability to move among other moving objects. Development of satellite navigation and electronic maps facilitated successful solution of the first task. The second task is much more difficult; many scientists around the world are trying to solve it using different methods and algorithms. The use of original path planning algorithms makes it easy to ensure the safety of movement in an area with dynamic obstacles.

2. Distribution of keypoints on the image

Any image represents complex and difficult-to-extract structured information about the observed scene. Therefore, we need a method that will allow extracting information from the real-time video data stream, identifying and recognizing objects using this information.

The main problem during the process of object recognition is to compare the image obtained from the camera with the etalon sample stored in the database. This problem can be solved using a number of tools allowing image recognition and matching them to databases. However, the main method is to establish a correspondence between the keypoints on the initial image and the etalon image.

A keypoint is the simplest geometric element of the discrete representation of the mathematical function for describing the object recognition. To define these points, we introduce a concept of a neighborhood. That is, we consider a point p_i to be a key (reference) point for some image; the neighborhood $O(p_i)$ of this point can be distinguished from the neighborhood $O'(p_i)$ of any other keypoint of the image p_i. The process of detecting the given point is called *detection*, and the program that carries out this process is called *a detector*. After this process, it is necessary to describe this particular point; and descriptors are responsible for this operation. A descriptor is a description of a keypoint that determines the characteristics of its neighborhood; it represents a numerical or binary vector of certain parameters [1].

There are many different methods for selecting keypoints and descriptors, but in this paper, we will use the histogram of oriented gradients (HOG). This choice is associated with a number of advantages of the HOG descriptor over others. First, HOG operates locally; it allows maintaining invariance in relation to geometric and photometric transformations of the object on small fragments of the image, but orientation of the object is an exception here. Second, a clear space partition, accurate calculation of directions, and strong local photometric normalization allow to not considering the movement of people if they are in vertical position. Therefore, this detector is a good tool for determining pedestrians on images.

3. Histogram of oriented gradients

The histogram of oriented gradients (HOG) means descriptors of keypoints used in computer vision and image processing systems for object recognition. This descriptor is based on the method of counting the number of gradient directions in local areas of the image.

The basis of this method is the assumption that an object description in the image can be achieved by specifying the distribution of intensity gradients or edge direction. The method is realized by partitioning the image into elementary regions (cells) and further calculating the histograms of the gradient directions or the edge directions for pixels for each region. The collection of histogram data for all cells is a descriptor [2].

Besides, the image is normalized in contrast; this is necessary in order to increase the accuracy. For this purpose, the intensity measure is determined for a large area of the image (block), the resulting value is used for normalization. This allows increasing the invariance of the descriptor in relation to illumination.

The steps of the HOG algorithm realization include the following: calculating the gradient, forming cell histograms, forming descriptor blocks, normalizing blocks, and classifying descriptors [3, 4]. The last step is the most difficult because it is based on machine learning. To implement it, the support vector method can be used. This method allows a binary classification, that is, a division into two classes: 1) the object belongs to the required category or 2) does not belong.

4. Supervised learning

The essence of the supervised learning is as follows. Let us suppose that we have a task to determine the membership of an object on the image. The decision-making process will be like a "black box" because we do not know how it works [5]. The task is to get a result or, in other words, to decide if an object belongs to a given group.

Let us consider the given process by an example of image recognition when it is necessary to define whether the given object is a person or not. In this case, we ourselves form the so-called "learning set", that is a set of examples and correct answers for them. We denote a set of examples as X, and a set of solutions as Y. Therefore, to solve our problem, it is necessary to introduce a function $f(X)$, which will allow transforming the set X into the set Y (Figure 1).

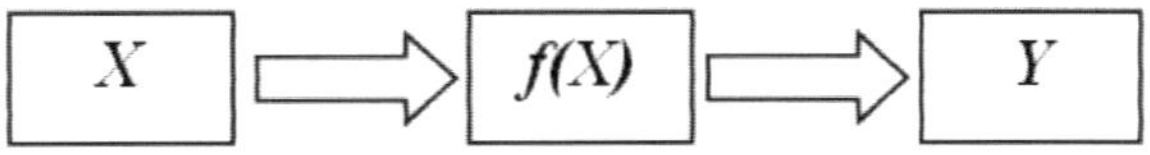

Figure 1. Function extraction $f(X)$.

These pairs of sets can be called as "stimulus-response" or precedent, and the aggregate of sets can be called as a training sample. The supervised learning is also often called as the learning by precedents [6].

Thus, the use of the function found and the software allows solving an example task, which was not presented in the training sample.

5. The support vector method

As we have already mentioned, the last step in the process of object detection on an image is the descriptor classification, which is implemented using the support vector method, which, in turn, is based on supervised learning.

The support vector method (SVM) is a set of algorithms for solving classification problems based on supervised learning; this method belongs to a family of linear classifiers. Its main advantage is in constant decrease in the empirical classification error and increase in the gap, that is, a more confident work of a classifier. Therefore, the method is also called as maximum margin classifier [7–9].

The main idea of the method is to interpret the initial vectors into a higher dimension space, and to allocate a hyperplane with the maximum gap in a given space.

5.1 Description of the support vector method using the optimal separating hyperplane

Let there be a linear separable sample, that is, a sample that can be divided into two classes (the case of binary classification). We divide this sample using a linear hyperplane; however, several planes may exist (Figure 2). In this case, each plane will divide the sample, but changes in the plane coefficients will be observed. Thus, here is a question. Which of these hyperplanes will be optimal?

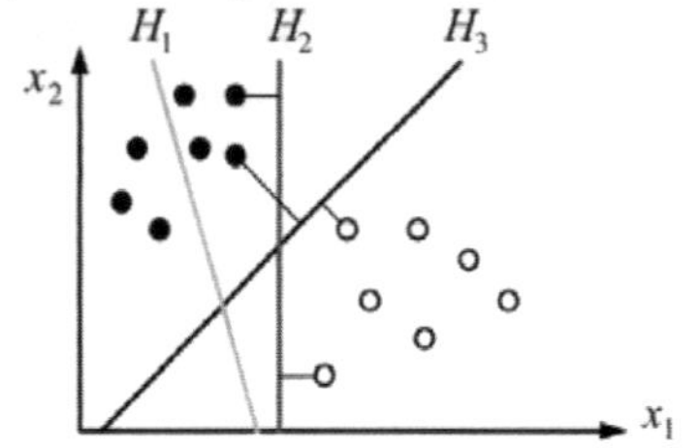

Figure 2. Construction of a separating hyperplane.

The optimal separating hyperplane will be the one that allows fulfilling the following condition: the distance between two nearest points lying on the opposite sides of the hyperplane (that is, between points belonging to different classes) should be maximum. The corresponding classifier is called as the optimal separating classifier.

To find the optimal hyperplane, we consider the formal construction, which is obtained as a result of the maximum shift of the separating hyperplane to to one or the other classes (Figure 3).

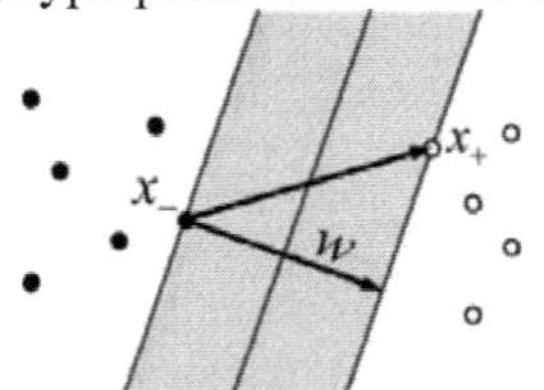

Figure 3. Construction of a separating hyperplane.

In this case, there is a question. What values will the scalar product of the weight vector w and x have taking into account the shear coefficient at the extreme points of each class. These extreme points belong to one or the other classes simultaneously, and the boundaries of the separating line pass through them.

$$a(x) = sign\left(\sum_{i=1}^{n} w_i x_i - w_0\right) = sign\left(\langle w, x\rangle - w_0\right), \langle w, x\rangle - w_0$$

where w_0 is a shear coefficient (the probability distribution parameter, it allows establishing a relationship between the value of this parameter and the choice of the reference point of the scale measurement); *sign* is a function that allows determining the sign of a number or a mathematical / trigonometric function.

If we multiply the vector w and the coefficient w_0 by some arbitrary number, then the separating surface will not change at all. The sign of the expression will not change because the function will only be multiplied by some constant. In this case, we can require that the values of this function on the extreme objects modulo equal to unity

$$\min_{i=1,\dots,l} y_i\left(\langle w, x_i\rangle - w_0\right) = 1 .$$

By requiring such normalization, it is possible to calculate the width of the separating strip. To do this, we take the direction that defines the normal to the separating surface, that is, the vector w, and consider the projection of the vector of extreme element difference of each class to this direction. In order to find this direction, it is necessary to take the scalar product of the *«x+»* and *«x-»* difference and the vector w normalized to its length. If we write this operation in details, we obtain the difference of scalar products of vectors:

$$\left\langle (x_+ - x_-), \frac{w}{\|w\|} \right\rangle = \frac{\langle w, x_+\rangle - \langle w, x_+\rangle}{\|w\|} .$$

At the same time, we should not forget that a certain relationship resulting from the normalization occurs to these extreme objects. If we use it, we get the following:

$$\left\langle (x_+ - x_-), \frac{w}{\|w\|} \right\rangle = \frac{\langle w, x_+\rangle - \langle w, x_+\rangle}{\|w\|} = \frac{(w_0 + 1) - (w_0 + 1)}{\|w\|} = \frac{2}{\|w\|}$$

As a result, the width of the separating strip will be equal to $2/\|w\|$. Now we can put down a task that will describe the maximization of the width of the separating strip. To do this, we minimize the scalar product of two vectors w taking into account the fact that the vector w length is the root of this value. We also should take into account that the indention on this object must be greater or equal to 1.

$$\begin{cases} \langle w, w\rangle \to \min \\ y_i\left(\langle w, x_i\rangle - w_0\right) \geq 1, \ i = 1, \dots l \end{cases}$$

5.2 Transition to linear inseparable sample

Let us consider the case of a linear inseparable sample. In the previous case, we could require that the indention was greater or equal to 1, where 1 was a positive value, and the indention was positive when the classification was correct. Now, in the case of a linear inseparable sample, there will be objects that are incorrectly assigned to the class by this classifier. This means that it is necessary to allow this algorithm to be mistaken, that is to allow making indention, which will not be greater or equal to 1, but will be greater than or equal to 1 minus ξ_i, where ξ_i is the error on the ith object. In this case, it is necessary to add "fines" for these errors to the function being minimized because in case of their absence, it is possible to make a classifier with any arbitrary indention

$$\begin{cases} \frac{1}{2}\langle w, w\rangle + C\sum_{i=1}^{l} \xi_i \to \min_{w, w_0, \xi} \\ y_i\left(\langle w, x_i\rangle - w_0\right) \geq 1 - \xi_i, \ i = 1, \dots, l \\ \xi_i \geq 0, \ i = 1, \dots, l \end{cases}$$

Thus, it is possible to obtain an optimization problem for the support vector method. We need to introduce a link to a linear classifier. First, it is necessary to minimize the sum ξ_i, with $\xi_i \geq 0$ and $\xi_i \geq$ 1- M_i, where M_i is an indention. This means that the error ξ_i will be greater or equal to the maximum of the two values - 0 and 1 minus indention. On the other hand, because the sum ξ_i is minimized, then ξ_i is exactly equal to this value, that is, the maximum of 0 and 1 minus M_i

$$\begin{aligned} &\xi_i \geq 0 \\ &\xi_i \geq 1 - M_i \quad \Rightarrow \xi_i = \max\{0, 1 - M_i\} = (1 - M_i)_+ \\ &\sum_{i=1}^{l} \xi_i \to \min \end{aligned}$$

In this case, you can simply substitute these ξ_i for the optimization task, namely for the first expression; then we will get an unconditional optimization problem in SVM. That is, an optimization task without additional conditions.

$$Q(w, w_0) = \sum_{i=1}^{l} \left(1 - M_i(w, w_0)\right)_+ + \frac{1}{2}\|w\|^2 \to \min_{w, w_0}$$

Here, we can clearly see the loss function, which is piecewise linear, and the regularizer, which is a regular L2-regularizer.

Thus, the support vector method is a linear classifier with a piecewise linear loss function (hinge loss) and an L2-regularizer. This method is necessary to maximize the gap between classes. In the case of a linear separable sample, this means simply maximizing the width of the separating strip. In the case of a linearly inseparable sample, the possibility of objects' hitting into the strip and "fines" for this hitting is simply added.

6. Conclusion

This algorithm, which includes a set of HOG descriptors and the support vector method, allows not only recognizing objects on a static image but also extracting them from the video data stream. The algorithm is optimal for recognizing people by means of an autonomous mobile object; it can also be used for recognizing other moving and static objects.

Acknowledgments

The research was carried out in Tomsk State University with the financial support of the Ministry of Education and Science of Russia, a unique project identifier: RFMEFI57817X0241.

The authors are grateful to Tatiana B. Rumyantseva from Tomsk State University for her assistance in preparing the paper.

References

[1] Finogeev A, Chetvertova M 2017 *Molodoj uchenyj* **11** 214–121
[2] Dalal N, Triggs B 2015 *Telecommunication Systems* **60** 337–339
[3] Shashev D, Shidlovsky S 2017 *Journal of Physics: Conference Series* **881(1)** 012029
[4] Hila M, Yannawar P, Gaikwad A 2017 *Proceedings - 1st International Conference on Intelligent Systems and Information Management, ICISIM 2017* 22–25
[5] Panin S V, Shakirov I V, Syryamkin V I, Svetlakov A A 2003 *Avtometriya (1)* 37–53
[6] Vyugin V 2014 *Matematicheskie osnovy mashinnogo obucheniya i prognozirovaniya* (Moscow, MCMNO)
[7] Gevan N, Ivanov V 2012 *Informacionnye tekhnologii i problemy matematicheskogo modelirovaniya slozhnyh sistem* **10** 81–84
[8] Wang Z, Qiao H 2017 *Proceedings of 2017 IEEE 7th International Conference on Electronics Information and Emergency Communication, ICEIEC 2017* 572–575
[9] Yurchenko A V *et al* 2015 *IOP Conf. Ser.: Mater. Sci. Eng.* **81** 012112

A numerical solution of the problem of crown forest fire initiation and spread

S I Marzaeva and O V Galtseva
Tomsk Polytechnic University, Tomsk, Russia
E-mail: rabota2013tpu@mail.ru

Abstract. Mathematical model of forest fire was based on an analysis of known experimental data and using concept and methods from reactive media mechanics. The study takes in to account the mutual interaction of the forest fires and three-dimensional atmosphere flows. The research is done by means of mathematical modeling of physical processes. It is based on numerical solution of Reynolds equations for chemical components and equations of energy conservation for gaseous and condensed phases. It is assumed that the forest during a forest fire can be modeled as a two-temperature multiphase non-deformable porous reactive medium. A discrete analog for the system of equations was obtained by means of the control volume method. The developed model of forest fire initiation and spreading would make it possible to obtain a detailed picture of the variation in the velocity, temperature and chemical species concentration fields with time. Mathematical model and the result of the calculation give an opportunity to evaluate critical conditions of the forest fire initiation and spread which allows applying the given model for of means for preventing fires.

1. Introduction

The characteristics of forest fires reveals their dependence on the specific conditions under which the experiments were conducted, and their insufficient accuracy, which results from the fact that under actual conditions it is impossible to control the meteorological situation and the homogeneity of forest fuel layer. For this reason, the physical modeling of forest fires must be combined with mathematical experiments using numerical methods and computers. Considering that, natural investigations of these problems are merely impossible. It was used the methods of mathematical modeling. Mathematical model of forest fire was based on an analysis of known experimental data and using concept and methods from reactive media mechanics. Usually crown fires are initiated by surface fires. The one of the first explanation of this process was given by Van Wagner [1, 2]. The theory proposed there depends on three simple crown properties: crown base height, bulk density and moisture content of forest fuel. Also crown fire initiation and hazard have been studied and modeled in detail [3–11]. The more complete mathematical model of forest fires was provided by Grishin [12–14]. In particular, this model which was obtained by Grishin [12] based on an analysis of known and original experimental data [2, 14], and using concepts and methods from reactive media mechanics. The physical multiphase model used in [15] may be considered as a continuation and extension of the formulation proposed by Grishin [12]. However, the investigation of crown fires has been limited mainly to cases studied of forest fires initiation without take into account the mutual interaction of the forest fires and three-dimensional atmosphere flows. This restriction has been removed in present setting of problem.

Boundary layer of the atmosphere, forest canopy and the lower tier of the forest is considered as a single region.

2. Physical and mathematical model

The basic assumptions using in this setting are: 1) the forest represents a multi-phase, multistoried, spatially heterogeneous medium; 2) in the fire zone the forest is a porous-dispersed, two-temperature, single-velocity, reactive medium; 3) the forest canopy is supposed to be non - deformed medium (trunks, large branches, small twigs and needles), affects only the magnitude of the force of resistance in the equation of conservation of momentum in the gas phase, i.e., the medium is assumed to be quasi-solid (almost non-deformable during wind gusts); 4) let there be a so-called "ventilated" forest massif, in which the volume of fractions of condensed forest fuel phases, consisting of dry organic matter, water in liquid state, solid pyrolysis products, and ash, can be neglected compared to the volume fraction of gas phase (components of air and gaseous pyrolysis products); 5) the flow has a developed turbulent nature and molecular transfer is neglected; 6) gaseous phase density doesn't depend on the pressure because of the low velocities of the flow in comparison with the velocity of the sound. Let the coordinate reference point x_1, x_2 , x_3= 0 be situated at the centre of the surface forest fire source at the height of the roughness level, axis $0x_1$ directed parallel to the horizontal surface to the right in the direction of the unperturbed wind speed, axis $0x_2$ directed perpendicular to $0x_1$ and axis $0x_3$ directed upward (Figure 1).

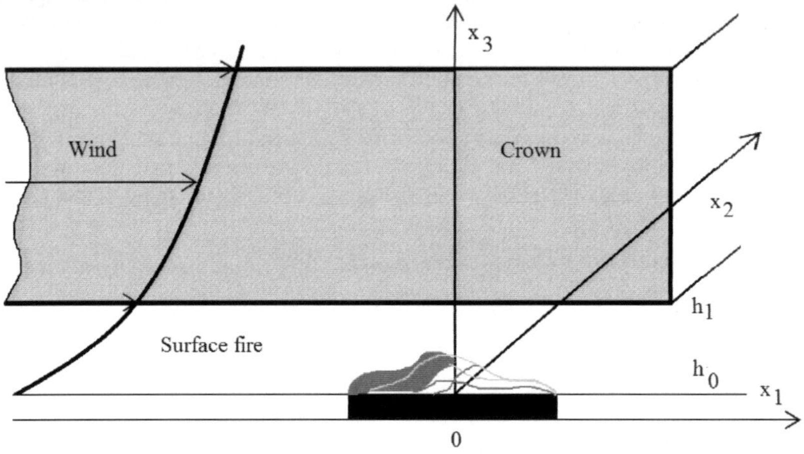

Figure 1. Computational domain.

Using the results of [12] and known experimental data [2, 12] governing equations are written as follows:

$$\frac{\partial \rho}{\partial t}+\frac{\partial}{\partial x_j}(\rho v_j)=\dot{m},\ j=1,2,\ i=1,2; \tag{1}$$

$$\rho\frac{dv_i}{dt}=-\frac{\partial P}{\partial x_j}+\frac{\partial}{\partial x_j}(-\rho\overline{v_i'v_j'})- \\ -\rho s c_d v_i|v|-\rho g_i-\dot{m}v_i; \tag{2}$$

$$\rho c_p\frac{dT}{dt}=\frac{\partial}{\partial x_j}(-\rho c_p\overline{v_j'T'})+q_5R_5-\alpha_v(T-T_s)+k_g(cU_R-4\sigma T^4); \tag{3}$$

$$\rho \frac{dc_\alpha}{dt} = \frac{\partial}{\partial x_j}(-\rho \overline{v'_j c'_\alpha}) + R_{5\alpha} - \dot{m} c_\alpha \;, \alpha = 1,5; \tag{4}$$

$$\frac{\partial}{\partial x_j}\left(\frac{c}{3k}\frac{\partial U_R}{\partial x_j}\right) - kcU_R + 4k_S \sigma T_S^4 + 4k_g \sigma T^4 = 0, k = k_g + k_S; \tag{5}$$

$$\sum_{i=1}^{4} \rho_i c_{pi} \varphi_i \frac{\partial T_S}{\partial t} = q_3 R_3 - q_2 R_2 + k_s (cU_R - 4\sigma T_S^4) + \alpha_V (T - T_S); \tag{6}$$

$$\rho_1 \frac{\partial \varphi_1}{\partial t} = -R_1, \rho_2 \frac{\partial \varphi_2}{\partial t} = -R_2, \rho_3 \frac{\partial \varphi_3}{\partial t} = \alpha_C R_1 - \frac{M_C}{M_1} R_3, \rho_4 \frac{\partial \varphi_4}{\partial t} = 0; \tag{7}$$

$$\sum_{\alpha=1}^{5} c_\alpha = 1,\; p_e = \rho RT \sum_{\alpha=1}^{5} \frac{c_\alpha}{M_\alpha}, \vec{v} = (v_1, v_2, v_3), \vec{g} = (0,0,g);\; \dot{m} = (1-\alpha_c) R_1 + R_2 + \frac{M_c}{M_1} R_3 + R_{54} + R_{55};$$

$$R_{51} = -R_3 - \frac{M_1}{2M_2} R_5, R_{52} = v_g (1-\alpha_c) R_1 - R_5, R_{53} = 0, R_{54} = \alpha_4 R_1, R_{55} = \frac{\alpha_5 v_3}{v_3 + v_{3*}} R_3.$$

The system of equation defines the state of the medium in the forest fire zone. Reaction rates of these various contributions (pyrolysis, evaporation, combustion of coke and volatile combustible products of pyrolysis) are approximated by Arrhenius laws whose parameters (pre-exponential constant k_i and activation energy E_i) are evaluated using data for mathematical models [12]. The boundary conditions for the system of equations (1)–(7) are specified as follows:

$$t = 0: v_1 = 0, v_2 = 0, v_3 = 0, T = T_e, c_\alpha = c_{\alpha e}, T_s = T_e, \varphi_1 = \varphi_{ie}; \tag{8}$$

$$x_1 = -x_{1e}: v_1 = V_e, v_2 = 0, \frac{\partial v_3}{\partial x_1} = 0, T = T_e, c_\alpha = c_{\alpha e}, -\frac{c}{3k}\frac{\partial U_R}{\partial x_1} + cU_R / 2 = 0; \tag{9}$$

$$x_1 = x_{1e}: \frac{\partial v_1}{\partial x_1} = 0, \frac{\partial v_2}{\partial x_1} = 0, \frac{\partial v_3}{\partial x_1} = 0, \frac{\partial c_\alpha}{\partial x_1} = 0, \frac{\partial T}{\partial x_1} = 0, \frac{c}{3k}\frac{\partial U_R}{\partial x_1} + \frac{c}{2} U_R = 0; \tag{10}$$

$$x_2 = x_{20}: \frac{\partial v_1}{\partial x_2} = 0, v_2 = 0, \frac{\partial v_3}{\partial x_2} = 0, \frac{\partial c_\alpha}{\partial x_2} = 0, \frac{\partial T}{\partial x_2} = 0, \frac{\partial U_R}{\partial x_2} = 0; \tag{11}$$

$$x_2 = x_{2e}: \frac{\partial v_1}{\partial x_2} = 0, \frac{\partial v_2}{\partial x_2} = 0, \frac{\partial v_3}{\partial x_2} = 0, \frac{\partial c_\alpha}{\partial x_2} = 0, \frac{\partial T}{\partial x_2} = 0, \frac{c}{3k}\frac{\partial U_R}{\partial x_2} + \frac{c}{2} U_R = 0; \tag{12}$$

$$\begin{aligned} &x_3 = 0: v_1 = 0, v_2 = 0, \frac{\partial c_\alpha}{\partial x_3} = 0, -\frac{c}{3k}\frac{\partial U_R}{\partial x_3} + \frac{c}{2} U_R = 0, v_3 = v_{30}, T = T_g, |x_1| \le \Delta, |x_2| \le \Delta, \\ &v_3 = 0,\; T = T_e, |x_1| > \Delta, |x_2| > \Delta; \end{aligned} \tag{13}$$

$$x_3 = x_{3e}: \frac{\partial v_1}{\partial x_3} = 0, \frac{\partial v_2}{\partial x_3} = 0, \frac{\partial v_3}{\partial x_3} = 0, \frac{\partial c_\alpha}{\partial x_3} = 0, \frac{\partial T}{\partial x_3} = 0, \frac{c}{3k}\frac{\partial U_R}{\partial x_3} + \frac{c}{2} U_R = 0. \tag{14}$$

Here and above $\frac{\mathrm{d}}{\mathrm{d}t}$ is the symbol of the total (substantial) derivative; α_V is the coefficient of phase exchange; ρ is density of gas – dispersed phase, t is time; v_i is the velocity components; T, T_S, are temperatures of gas and solid phases, U_R is density of radiation energy, k *is* coefficient of radiation attenuation, P is pressure; c_p is constant pressure specific heat of the gas phase, c_{pi}, ρ_i, φ_i are specific heat, density and volume of fraction of condensed phase (1 is dry organic substance, 2 is moisture, 3 are condensed pyrolysis products, 4 is mineral part of forest fuel), R_i are the mass rates of chemical reactions, q_i are thermal effects of chemical reactions; k_g , k_S are radiation absorption coefficients for gas and condensed phases; T_e is the ambient temperature; c_α are mass concentrations of α - component of gas - dispersed medium, index α=1,2,...,5, where 1 corresponds to the density of oxygen, 2 - to carbon monoxide CO, 3 - to carbon dioxide and inert components of air, 4 - to particles of black, 5 - to particles of smoke; R is universal gas constant; M_α , M_C, and M are molecular mass of α -components of the gas phase, carbon and air mixture; g is the gravity acceleration; c_d is an empirical coefficient of the resistance of the vegetation, s is the specific surface of the forest fuel in the given forest stratum, g is mass fraction of gas combustible products of pyrolysis, α_4 and α_5 are empirical constants. To define source terms which characterize inflow (outflow of mass) in a volume unit of the gas-dispersed phase were used the following formulae for the rate of formulation of the gas-dispersed mixture $\dot{m}$, outflow of oxygen R_{51} , changing carbon monoxide R_{52} , generation of black R_{54} and smoke particles R_{55} .

Coefficients of multiphase (gas and solid phase) heat and mass exchange are defined $\alpha_V = \alpha S - \gamma C_p \dot{m}, S = 4\phi_S / d_S$. Here $\alpha = Nu\lambda/d_S$ is coefficient of heat exchange for sample of forest combustible material (for example needle), Nu is Nusselt number for cylinder, λ is coefficient of heat conductivity for pine needle; γ is parameter, which characterize relation between molecular masses of ambient and inflow gases. It is supposed that the optical properties of a medium are independent of radiation wavelength (the assumption that the medium is “grey”), and the so-called diffusion approximation for radiation flux density were used for a mathematical description of radiation transport during forest fires. The components of the tensor of turbulent stresses, as well as the turbulent fluxes of heat and mass are written in terms of the gradients of the average flow properties [12]. It should be noted that this system of equations describes processes of transfer within the entire region of the forest massif, which includes the space between the underlying surface and the base of the forest canopy, the forest canopy and the space above it, while the appropriate components of the data base are used to calculate the specific properties of the various forest strata and the near-ground layer of atmosphere. The thermodynamic, thermophysical and structural characteristics correspond to the forest fuels in the canopy of a different type of forest (for example, pine forest [12-16]).

3. Numerical solution and results

The boundary-value problem (1)–(14) was solved numerically using control volume method and the SIMPLE like algorithm [16]. Also numerical calculation carried out using software PHOENICS [17, 18]. Fields of temperature, velocity, component mass fractions, and volume fractions of phases were obtained for different instant of time. Figures 2*a-g* present the distribution for temperature and velocities.

Figures 2*a-g* present the distribution for temperature (*a-t=7s, b-t=9s, c – t=14s, d- 16s, e – t=17s, f – t=21s, g – t=27 s.*) and velocities. The isothermes are moved in the forest canopy and deformed by the action of wind. The results of calculation give an opportunity to evaluate critical condition of the forest fire spread, which allows applying the given model for preventing fires. It overestimates the crown forest fire spread that depends on crown properties: bulk density, moisture content of forest fuel and etc. The model proposed there gives a detailed picture of the distribution in the velocity, temperature and component concentration fields with time, and determine as well as the influence of different conditions on the crown forest fire spread.

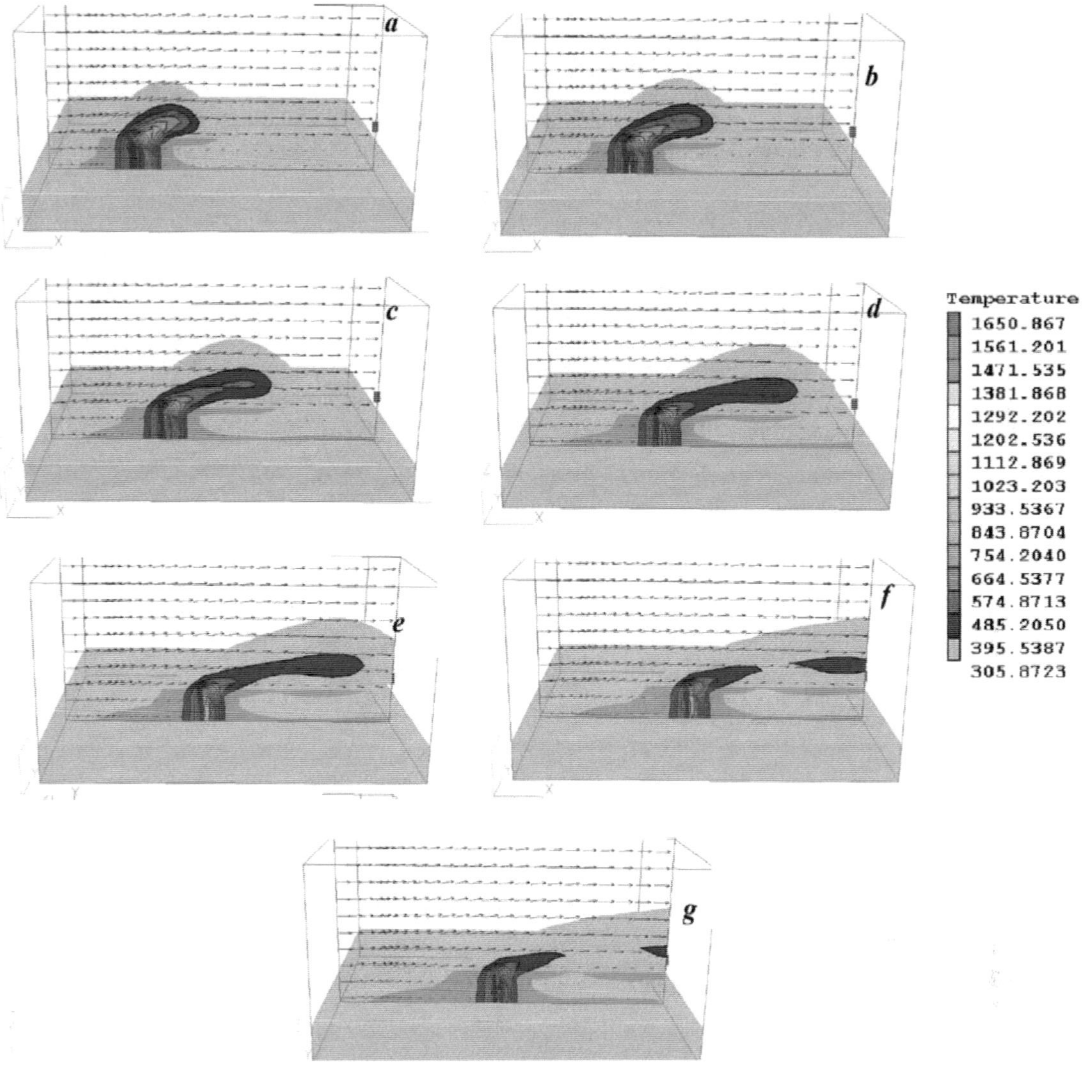

Figure 2. The distribution of temperature and field of velocity.

4. Conclusion

A multiphase mathematical model of crown fire initiation and spread was developed using the multi-purpose software, PHOENICS [17, 18]. The model accounts for all the important physical and physicochemical processes: drying, pyrolysis, char combustion, turbulent combustion of gaseous products of pyrolysis, exchange of mass, momentum and energy between gas and solid phase, turbulent gas flow and convective, conductive and radiative heat transfer.

References

[1] Van Wagner C E 1977 *Can J For Res* **7(1)** 23–34 doi: 10.1139/x77-004
[2] Van Wagner C E 1979 *Proc of the X Conference on Fire and Forest Meteorology: Ottawa, Ontario* 207–212
[3] Konev E V 1977 *Mathematical Modelling Forest Fire and New Methods Fighting Them The*

physical foundation of vegetative materials combustion (Novosibirsk, Nauka)
[4] Xanthopoulos G 1990 Development of a wildland crown fire initiation model *PhD Thesis* University of Montana: USA, 152 p.
[5] Rothermel R C 1991 *Proc of the 11th Conference on Fire and Forest Meteorology: Missoula, Montana* 253–263
[6] Rothermel R C 1988 *J Wildland Fire* **1** 1–10
[7] Van Wagner C E 1993 *Can J For Res* **23(3)** 442–449 doi: 10.1139/x93-062
[8] Cruz M G et al 2002 *Proc of the IV Conference on Fire and Forest* Meteorology: *Luso-Coimbra, Portugal* 1–11 ISBN: 90-77017-72-0
[9] Albini F A *et al* 1995 *Int J Wildland Fire* **5** 81–91 doi: 10.1071/WF9950081
[10] Scott J H *and* Reinhardt E D 2001 *USDA Forest Service, Rocky Mountain Research Station,* RMRS-RP-29
[11] Bordunov S V *et al* 2018 *IOP Conf. Ser.: Mater. Sci. Eng.* **289(1)** 012002 doi: 10.1088/1757-899X/289/1/012002
[12] Grishin A M 1997 *Mathematical Modelling Forest Fire and New Methods Fighting Them* (Tomsk, TSU Publishing House)
[13] Grishin A M *and* Perminov V A 1996 *Fiz Gorenia Vzryva* **32(5)** 107–115
[14] Grishin A M *and* Perminov V A 1993 *Fiz Goreniya Vzryva* **29(6)** 8–14
[15] Morvan D *and* Dupuy J L 2004 *Combust. Flame* **138** 199–210
[16] Patankar S V 1981 *Numerical Heat Transfer and Fluid Flow* (Hemisphere Publishing Corporation, New York)
[17] *PHOENICS On-Line Information System* Retrieved from http://www.cham.co.uk/PHOENICS/d_polis/polis.htm
[18] Agranat V *et al* 2015 *Proc of ICONE'2015 Chiba-city: Japan*

Mathematical model of solar radiation based on climatological data from NASA SSE

S G Obukhov[1], I A Plotnikov[1], V G Masolov[2]
[1]National Research Tomsk Polytechnic University, Tomsk, Russia
[2]"VDM-tekhnika" LLC, Moscow,Russia
E-mail: serob99@mail.ru

Abstract. An original model of solar radiation arriving at the arbitrarily oriented surface has been developed. The peculiarity of the model is that it uses numerical values of the atmospheric transparency index and the surface albedo from the NASA SSE database as initial data. The model is developed in the MatLab/Simulink environment to predict the main characteristics of solar radiation for any geographical point in Russia, including those for territories with no regular actinometric observations.

1 Introduction

A reliable prediction of solar radiation arriving at the proposed plant site is crucial to design a solar power plant. For many years, the USSR climate reference books, which contain data on hourly, monthly and annual values of direct, scattered and total solar radiation, and sunshine duration in all regions of Russia for observation periods from 5 to 30 years, were the main source of actinometric data required for solar energy calculations [1, 2]. However, a limited edition and inconvenient form of data representation significantly limit the possibilities of the data practical application. A small number of meteorological stations that perform regular actinometric observations (no more than 130 [3]) considerably complicates obtaining reliable data on the intensity of solar radiation for most territories of Russia.

The list of the world and European electronic databases represented in a more convenient form and their main characteristics are provided in [4, 5]. The results of satellite measurements and the models of solar radiation propagation in the atmosphere are used for creating up-to-date databases to enable interpolation of the data on the intensity of solar radiation over large areas. Disadvantages of most foreign databases are commercial access, limited coverage of the territory, and as a result, the lack of data for most regions of Russia. Russian experience in creating electronic climate databases is not great. At present, data on actinometric observations for a limited number of stations included in the World Meteorological Organization (WMO) are available in the archive [6] of the World Radiation Data Center and on the official website of All-Russian Scientific Research Institute of Hydrometeorological Information [7].

Atlas of solar energy resources in Russia [3] developed at Joint Institute for High Temperatures RAS is of particular interest. The Atlas provides interactive maps of the distribution of the daily total and direct solar radiation arriving at a fixed surface tilted at different angles to the horizon and at the surface tracking the sun across the territory of Russia. The Atlas is available on the website of the project Geoinformation System, Renewable Energy Sources of Russia [8], and provides the designers of solar power plants with the required initial data to assess the potential of solar energy in various geographical regions of Russia.

However, in spite of the fact that a generalized technical and economic analysis of the efficiency of solar power plants can be performed using electronic databases, the data provided in databases are not enough to solve the majority of problems in technical design. The problems include the determination of the optimal ratio of capacities between the main and reserve generating sources, the required capacity for storage devices, the efficient tracking system and its parameters, the analysis of the energy balance in the designed system, etc. These problems are particularly relevant to hybrid power plants with renewable energy sources designed for autonomous consumer. The commensurability of generating power and load capacities requires the analysis of the plant operating conditions and the concurrence between the production and power consumption modes for both long (month, season) and relatively short (hours, days) time intervals. The stochastic nature of renewable energy sources significantly complicates the solution of the above problems and causes the necessity to use imitative mathematical modeling methods in designing similar power systems. Important components of the imitative models of power systems with renewable energy sources are the models of primary energy resources that contribute to their modeling with high time sampling.

The present study aims to develop a mathematical model of solar radiation arriving at an arbitrarily oriented surface located in any region of Russia.

2 Review of mathematical models of solar radiation

The value of solar radiation arriving at the receiving surface of solar panels depends on a number of factors, some of which are deterministic, and some are stochastic in nature. The deterministic factors include geographical latitude of the solar power plant location, day of year and time of day, and orientation angles of the receiving surface of the solar panel relative to the Sun. Stochastic factors that largely determine the intensity of solar radiation include the concentration of atmospheric gases, dust, aerosols and water vapor suspended in the air, and the nature of cloud cover and underlying surface.

A detailed review, classification and comparative analysis of the solar radiation models are presented in [9–13]. In most models, deterministic factors are determined using the known analytical equations established by solar geometry, and stochastic factors are determined on the basis of statistical processing of the results of long actinometric observations.

The transparency coefficient K_T and the diffusion coefficient K_D that show the intensity of solar radiation arriving at the horizontal surface are most often used as the integral characteristics of the state of the atmosphere:

$$K_T = \frac{H}{H_0}, \quad K_D = \frac{H_D}{H} \tag{1}$$

where H and H_D are the average daily intensity of total and scattered solar radiation arriving at the horizontal surface, respectively; H_0 is the intensity of solar radiation at the upper boundary of the atmosphere.

Numerous studies have shown a close correlation between the diffuse coefficient K_D and the transparency coefficient K_T, which can be represented as an empirical equation:

$$K_D = f(K_T) \tag{2}$$

Polynomial equations are often used to describe the dependence (2), the most common of which are Erbs's model [14], Orgilland Hollands's model [15], Reindl's model [16], etc. A detailed review of the models of this type is reported in [11].

This study employs the NASA methodology [17] to determine the value of the diffuse coefficient K_D. According to the methodology, the fraction of scattered radiation is determined from the system of polynomial equations as a function of the latitude of the locality φ, the transparency coefficient K_T, the sunset hour angle ω_{ss}, and the midpoint elevation angle of the Sun above the horizon h.

There are two main groups of models to convert the scattered radiation from a horizontal surface to a tilted one: isotropic and anisotropic. The isotropic models (Liu and Jordan's [18], Koronakis' [19] et al.) assume a homogeneous distribution of scattered radiation in the sky. The anisotropic models

(Klucher's [20], Hay's [21] et al.) consider the relationship between the direction of scattered radiation propagation and the zenith angle of the Sun.

In the present study, anisotropic Hay's model is used to determine the values of scattered radiation [21]. According to the results of the studies presented in [22], the model provides higher accuracy in calculating the values of scattered radiation in high northern latitudes.

3 Theoretical description of the proposed model

The study uses a combined model of the solar radiation arrival, in which some of the parameters are calculated from analytical equations, and some are determined using empirical coefficients from the NASA SSE database [23] for the specified solar power plant location. The total solar radiation arriving at the tilted surface is determined by the equation:

$$G = (G_H - G_{DH}) \cdot \frac{\cos\theta}{\cos\theta_z} + G_{DH} \cdot \left[A_i \cdot \frac{\cos\theta}{\cos\theta_z} + (1 - A_i) \cdot \left(\frac{1+\cos\beta}{2} \right) \right] + G_H \cdot \rho \cdot \frac{1-\cos\beta}{2}, \qquad (3)$$

where G_H and G_{DH} are the values of total and scattered solar radiation arriving at the horizontal surface, respectively; θ is the angle between the incident radiation direction to the surface and the normal to the surface; θ_z is the solar zenith angle; ρ is the Earth's surface albedo; β is the inclination angle of the receiving surface to the horizontal surface; A_i is the anisotropy index determined by the equation:

$$A_i = \frac{G_H - G_{DH}}{G_0}, \qquad (4)$$

where G_0 is extra-atmospheric solar radiation on the horizontal surface.

The value of the angle θ is determined by formula:

$$\begin{aligned} \cos\theta = {} & \sin\delta \cdot \sin\varphi \cdot \cos\beta - \sin\delta \cdot \cos\varphi \cdot \sin\beta \cdot \cos\gamma + \cos\delta \cdot \cos\varphi \cdot \cos\beta \cdot \cos\omega + \\ & + \cos\delta \cdot \sin\varphi \cdot \sin\beta \cdot \cos\gamma \cdot \cos\omega + \cos\delta \cdot \sin\beta \cdot \sin\gamma \cdot \sin\omega, \end{aligned} \qquad (5)$$

where φ is the latitude of the receiving surface locality; δ is the solar declination angle; γ is the azimuth angle of the receiving surface (with a south-facing direction γ=0°, with a deviation to the east γ is considered negative, with a deviation to the west γ is positive); ω is the solar hour angle.

The declination angle is calculated by the formula:

$$\delta = 23.45 \cdot \sin\left[\frac{360}{365} \cdot (N + 284) \right], \text{ deg.} \qquad (6)$$

where N is the day number, such that N=1 on the 1st January.

The solar zenithal angle is calculated by the formula:

$$\theta_z = \arccos[\sin\varphi \cdot \sin\delta + \cos\varphi \cdot \cos\delta \cdot \cos\omega], \text{ deg.} \qquad (7)$$

The zenithal angle is the conjugate angle of the solar elevation above the horizon h, which can be calculated from:

$$h = 90° - \theta_z \qquad (8)$$

The solar azimuth angle Az can be calculated from the following equation:

$$\cos Az = \frac{\sinh \cdot \sin\varphi - \sin\delta}{\cosh \cdot \cos\varphi} \qquad (9)$$

The azimuth angles at sunrise ω_{sr} and sunset ω_{ss} can be calculated from:

$$\begin{aligned} \omega_{sr} &= 0 - \arccos[-\text{tg}\varphi \cdot \text{tg}\delta] \\ \omega_{ss} &= 0 + \arccos[-\text{tg}\varphi \cdot \text{tg}\delta] \end{aligned} \qquad (10)$$

The average daily values of extra-atmospheric radiation arriving at the horizontal surface are determined by the equation:

$$H_0 = \frac{24}{\pi} \cdot G_{sc} \cdot \left(1 + 0.033 \cdot \cos\frac{360 \cdot N}{365} \right) \cdot (\cos\varphi \cdot \cos\delta \cdot \sin\omega_s + \sin\varphi \cdot \sin\delta), \qquad (11)$$

where G_{sc}=1 367 Wt/m^2 is the solar constant.

The current daily values of total and scattered solar radiation on the horizontal surface are calculated by the formulas:

$$G_H = r_t \cdot H, \qquad G_{DH} = r_d \cdot H_D \tag{12}$$

To calculate the coefficients from equation (12), we use the formulas proposed by Collares-Pereira [24]

$$r_t = \frac{\pi}{24} \cdot (a + b \cdot \cos\omega) \cdot \frac{\cos\varphi - \cos\omega_s}{\sin\omega_s - \omega_s \cdot \cos\omega_s}, \tag{13}$$

with coefficients

$$\begin{aligned} a &= 0.409 + 0.5016 \cdot \sin(\omega_s - 60) \\ b &= 0.6609 - 0.4767 \cdot \sin(\omega_s - 60), \end{aligned} \tag{14}$$

and by Liu and Jordan [17]:

$$r_d = \frac{\pi}{24} \cdot \frac{\cos\varphi - \cos\omega_s}{\sin\omega_s - \omega_s \cdot \cos\omega_s} \tag{15}$$

To determine the real characteristics of solar radiation arriving at a certain territory, the model uses the numerical values of the transparency coefficient K_T and the surface albedo ρ from the NASA SSE database [23]. Advantages of this database are open access, convenient service and up-to-date models of the solar radiation propagation in the atmosphere that take into account the features of the climate zones, the nature of radiation reflected from the Earth's surface, cloud amount, and atmospheric pollution. The NASA database is based on the ground and satellite observations of solar radiation in the period from 1983 to 2005, and the calculated values of solar radiation are interpolated to a 1°×1° grid covering the Earth. The comparative analysis of the NASA database data with the results of direct long-term measurements of solar radiation from 50 Russian meteorological stations located in different regions of the country showed that the average radiation error for most regions of Russia does not exceed 15% [5].

Based on equations (1)–(15), a mathematical model was developed in the MatLab/Simulink environment to determine the amount of solar radiation arriving at the arbitrarily oriented surface for any day of the year in any location of the solar power plant.

The model verification using the observation data from 10 meteorological stations located in the territory 43° to 62° N, 50° to 158° E showed that the model provides a satisfactory error in modeling both the annual and daily course of solar radiation. The average relative error for all the cities under consideration does not exceed 11.7% for total solar radiation and 24.5% for scattered radiation [25], which is quite acceptable for engineering calculations.

4 Results and discussion

Illustrate the possibilities of the proposed model through a specific example of determining the potential solar energy resources for the solar power plant located in Tomsk region (56.5° N, 85° E).

Table 1 summarizes the average daily values of total solar radiation on the horizontal surface under actual cloud conditions, according to the data from the USSR Scientific and Applied Reference Book on Climate [1], the NASA SSE database [23] and the results of modeling.

Table 1. The average daily total solar radiation on the horizontal surface under actual cloud conditions (kWh/m^2) in Tomsk

	Jan.	Feb.	Mar.	Apr.	May	June	July	Aug.	Sept.	Oct.	Nov.	Dec.	year
Reference book	0.67	1.54	3.15	4.27	4.99	5.70	5.48	4.20	2.87	1.32	0.74	0.45	35.39
NASA SSE	0.69	1.59	2.94	4.29	5.48	5.79	5.80	4.55	2.83	1.58	0.83	0.46	36.83
Model	0.66	1.53	2.90	4.17	5.43	5.79	5.84	4.48	2.72	1.44	0.77	0.45	36.18

While modeling, the "average day of the month" method proposed by Klein [26] was used to reduce the amount of calculation. According to the method, the daily average intensity of solar radiation for the considered month of the year is equal to the daily radiation intensity for the corresponding day. The analysis of the results presented in Table 1 shows that the model provides satisfactory accuracy in calculating the total values of solar radiation arriving at the horizontal surface: the relative error for all months of the year is no more than 10%.

For the comparative evaluation of the efficiency of solar tracking systems in Tomsk, a series of computational experiments was performed to determine the values of solar radiation arriving at the receiving surface of satellites oriented at different angles. As an example, Figure 1 illustrates the calculated dependences of solar radiation arriving at the surface of the solar panel on the day of the summer solstice.

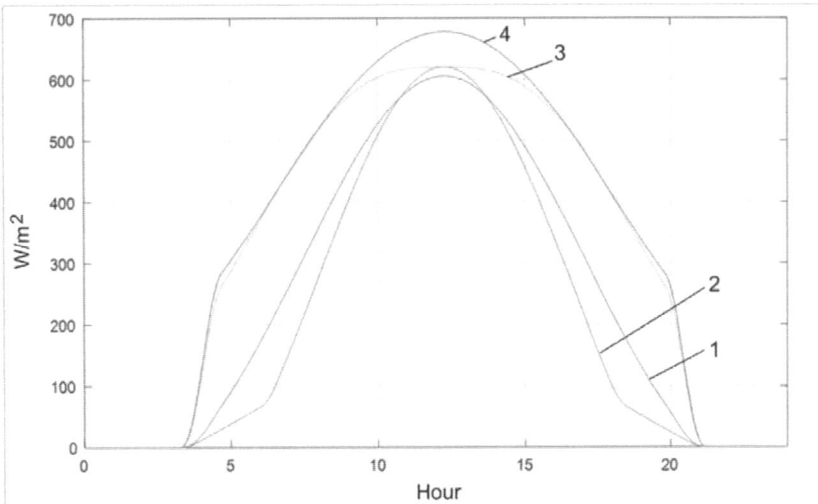

Figure 1. Dependencies of total solar radiation arriving at the surface of the solar panel on the day of the summer solstice. The solar panel is installed horizontally (1), at the angle of 56.5° to the horizon (2), with a single-axis (3) and two-axis tracking systems (4).

The analysis of modeling results shows that the proposed model adequately reflects the daily course of solar radiation and can be used as a convenient tool for solving many practical problems associated with the design and optimization of the power systems, which employ solar radiation as a primary energy source.

5 Conclusions

The original model of solar radiation arriving at the arbitrarily oriented surface has been developed. The peculiarity of the model is that it uses numerical values of the atmospheric transparency index and the surface albedo from the NASA SSE database as the initial data. This enables using the model to predict the main characteristics of solar radiation for any geographical point in Russia, including those for territories with no regular actinometric observations.

Acknowledgments

The study was supported by the Ministry of Education and Science of the Russian Federation. The unique identifier is RFMEFI57617X0098 (agreement No. 14.576.21.0098 dated 26 September 2017).

References

[1] 1967 *Reference book on the climate of the USSR. Solar radiation, radiation balance and sunshine* (Leningrad,Gidrometeoizdat)
[2] 1993 *Scientific and Applied Reference Book on the Climate of the USSR. Series 3. Long-term data* (St.P., Gidrometeoizdat)
[3] Popel O S, Frid S E, Kolomiets Yu G, Kiseleva S V 2010 (Moscow, MFT Publishing House)
[4] Vernaya C, Pitaval S, Blanc P 2014 *Energy Proceedings* **57** 1383–1391
[5] Popel O S *et al* 2009 *Climatic data for renewable energy in Russia (climate database): textbook* (Moscow, MIPT Publishing House)
[6] World Radiation Data Centre. URL: http://wrdc.mgo.rssi.ru/
[7] All-Russia Research Institute of Hydrometeorological Information. URL:http://meteo.ru/
[8] Geoinformation system "Renewable energy sources of Russia". URL: http://gisre.ru/
[9] Zhang J *et al.* 2017 *Renewable and Sustainable Energy Reviews* **70** 314–329
[10] Besharatetal F, Dehghan A A, Faghih A R 2013 *Renewable and Sustainable Energy Reviews* 798–821
[11] Maleki S A M, Hizam H, Gomes C 2017 *Energies* **10(1)** 134; doi:10.3390/en10010134
[12] Katiyar A K, Pandey C K 2013 *Hindawi Publishing Corporation Journal of Renewable Energy* **2013** 168048 doi:10.1155/2013/168048
[13] Petrusev A S et al 2016 2nd International Conference on Industrial Engineering, Applications and Manufacturing, ICIEAM 2016 7911461 doi: 10.1109/ICIEAM.2016.7911461
[14] Erbs D G, Klein S A, Duffie J A 1982 *Solar Energy* **28(4)** 293–302
[15] Orgill J K, Hollands G T 1977 *Solar Energy* **19(4)** 357–359
[16] Reindl D T, Beckman W A, Duffie J A 1990 *Solar Energy* **45(1)** 1–7
[17] Stackhouse P W *et al* 2016 Surface meteorology and Solar Energy (SSE) Release 6.0 Methodology, Version 3.2.0 June 2,. URL: https://eosweb.larc.nasa.gov/cgi-bin/sse/sse.cgi?skip@larc.nasa.gov+s07#s07
[18] Liu B Y H, Jordan R C 1961 *ASHRAE* **10** 53–59
[19] Koronakis P S 1986 *Solar Energy* **36(3)** 217–225
[20] Klucher T M 1979 *Solar Energy* **23(2)** 111–114
[21] Hay J E 1979 *Solar Energy* **23(4)** 301–307
[22] Stadnik V V, Razgonyaev Y V *Proceedings of the Geophysical Observatory named after A.I. Voeikov* 557 67–84
[23] The NASA Surface Meteorology and Solar Energy Data Set. URL: http://eosweb.larc.nasa.gov/sse/
[24] Collares-Pereira M, Rabl A 1979 Solar Energy **22** 155–164
[25] Obukhov S G, Plotnikov I A 2017 *International Scientific Journal. Alternative Energy and Ecology* **(16–18 (228–230))** 43–56
[26] Klein S A 1977 Solar Energy **19(4)** 325–329

Fundamentals of thinking, patterns

O M Gafurov, D O Gafurov, V I Syryamkin
National Research Tomsk State University, Tomsk, Russia
E-mail: oleg.gafurov@mail.ru

Abstract. The authors analyze the fundamentals of thinking and propose to consider a model of the brain based on the presence of magnetic properties of gliacytes (Schwann cells) because of their oxygen saturation (oxygen has paramagnetic properties). The authors also propose to take into account the motion of electrical discharges through synapses causing electric and magnetic fields as well as additional effects such as paramagnetic resonance, which allows combining multisensory object-related information located in different parts of the brain. Therefore, the events of the surrounding world are reflected and remembered in the cortex columns, thus, creating isolated subnets with altered magnetic properties (patterns) and subsequently participate in recognition of objects, form a memory, and so on. The possibilities for the pattern-based thinking are based on the practical experience of applying methods and technologies of artificial neural networks in the form of a neuroemulator and neuromorphic computing devices.

1 Introduction

The human brain also has evolved like the human body with its sense organs. The brain belongs to our planet Earth at least in its basic ancient parts, appears, and changes in its embryonic development repeating the evolution. Its change is closely related to the change of the human body. Simultaneously with the development of the embryo, autonomous self-regulated control systems with the basic life-supporting functions of the body appear in brain regions.

Unconditioned reflexes attributed to the human cognitive functions are embedded in a human being as a necessary and sufficient condition for survival and development of the organism on the surface of the planet Earth. If necessary, in case of stopping development of the embryo on the third week, apparently from the same material, we can get amphibians able to live and breathe under water. In embryonic development, the more ancient parts of the human brain (oblong, intermediate, dorsal and others) are formed accordingly; they provide vital functions for the human activity control regardless of the cogitative activity (a person in a coma with no signs of cognitive functions continues to exist, and in case of appropriate care he breathes, is nourished, etc.).

In full understanding, cognitive thought processes are manifested in the frontal lobes. Cognitive functions and the thought process are provided by formation and training of neural networks (patterns) of the cerebral cortex, which are cortex columns (subnets) consisting of neurons, synapses and gliacytes.

If there is a sufficiently strong excitation in one of the brain regions, an acoustic signal appears and the local neuron networks created perceive it with the same natural frequency. It means that the pattern is excited and it delivers the image recognized to the receptors of the second signaling system, i.e. we perceive the known object all over again.

Pattern-based thinking, as will be shown based on the realized and approved systems, allows using almost the same program by implementing it in the form of a neuroemulator or neurochip, because the brain is also divided into isolated regions responsible for various functions of recognition and formation of an integrated solution.

Problems of acoustic and seismoacoustic analysis of encephalograms and cardiograms (solution for wave control processes) were realized in the artificial neural network *NeuroCiber*. Problems of analysis of images, geophysical and geochemical surface measurements (geomagnetic, gravity measurements, aerial photography, etc.) as well as problems of analysis of large data sets represented in the form of two and three-dimensional data were realized in the artificial neural network *NeuroInformGeo* (the patent available). Problems of analysis and forecast of economic, social, and political data as well as analysis and forecast of time series were realized in the artificial neural network *NeuroFinExpert*.

Realization of the proposed algorithms for the general solution of the wave equation by means of neural networks with creation of phoneme and ence phoneme patterns and patterns of various diseases of the brain and heart allows us to solve a wide range of problems including recognition of mentally spoken phrases and their direct and mental transmission. We control artificial *NeuroCiber* objects by voice commands [1–3].

2 Human brain

The brain is millions of years old, but at the same time, it has rather young formations. Formation of the parts of the human brain occurred at different evolution periods and each of them was an independent self-regulating system. Let us distinguish the basic brain formations: the cerebral cortex, medulla oblongata, and interbrain. The interbrain (hypothalamus) contains the regions (volumes) of the nerve centers responsible for hunger and saturation. The medulla oblongata contains the breathing regulation apparatus.

3 Fundamentals of thinking

Neurophysiologists formulated the law of a nervous impulse transmission across synapses, identified visual and auditory zones, constructed a map of projections of thalamic nuclei on the cerebral cortex, and others. However, one of the most profound unsolved scientific problems is the nature of consciousness. How is consciousness generated by the brain?

The general structure of consciousness consists of **visual, auditory, tactile, and somatic** zones. All this zones are superimposed in a **living neural network**. It means that the method of calculation lies in the trained and constantly existing neural networks (when creating neural robots or at birth of an animal or a human being) and these neural networks form a system of unconditioned reflexes. They must ensure the survival of a living organism or a neuro robot in the real physical and temporal continuum (for a human being it is the planet Earth of the solar system in our Galaxy). Additional skills are formed in the same system, which means that the new neural networks are formed in other parts of the brain (conditioned reflex).

The basic paradigm is that only the following libraries of neural networks participate in all calculations: ready, inherited, formed at the previous stage or by previous generations of neural robots (as in our case), and neural networks trained during the life of living creatures or during the functioning of a neural robot in the real-time mode (patterns).

4 The basic concepts of brain activity of living creatures

1. Inherited (unconditioned) reflexes are obtained mainly during the evolutionary development of our organism; they ensure connection of the organism with a given planet and its daylight surface, and with an opportunity to return to the Ocean.
2. Conditioned reflexes are the basis for training, after-training and retraining of neural networks.
3. The function of pattern recognition is a fundamental property of a living organism. A fundamentally important feature of pattern recognition by a neural network is the ability to reconstruct an image from the data reduced, distorted or incomplete, and stored in the memory.

For example, a simplified image or its part is to be recognized and the task of the neural network is to restore the complete initial image stored in the memory.
4. Modelling of external processes and events (sleep is a period for cleaning, simultaneous and constant repetition of events already happened, highlighting the most significant ones and their memorizing, retraining of the networks already retrained).
5. Combination of training, after-training, and retraining with recognition and modeling of physical events and processes, which have really happened, leads to the appearance of conditioned reflexes and their consolidation, i.e. appearance of skills (for living creatures, it is usually associated with emotional experience reflected in a certain region on the brain).
6. Chaotic dynamics of the magnetic and electrical activity of the brain. Studies have shown that the electroencephalograms (EEG) of animals and humans, as well as magnetic fields generated in the interneuronic brain tissues, represent deterministic chaotic processes with a small number of degrees of freedom.
7. Parallel methods of processing of incoming and simulated information in the brain.
In this paper, the authors propose to consider a model of the brain, which includes the following aspects:
8. Magnetic properties of gliacytes (Schwann cells).
9. Subdivisions of the neural network structure of the brain, cortex columns (subnets).
10. Oscillation of various parts of the brain, resonance and feeding of the processed signals again to the first signaling system into the visual, auditory, and somatic zones in full or partial volume.
These elements are necessary and sufficient conditions for constructing human thinking or thinking of another living creature based on patterns (experience accumulated in the previous life), which are both memory and means of recognition of objects and events [4].

5 Sense organs

Sense organs give us the world full of colors, sounds and smells.

Sense organs are analyzers that signal to receptors, i.e. cells that process and transform signals into nerve impulses.

Sense organs represent a specialized peripheral anatomical and physiological system, which provides (by means of its receptors) receipt and primary analysis of information from the environment and from other organism's organs, that is, from the organism's external and internal environment.

Remote sense organs perceive stimuli at a distance (for example, organs of sight, hearing, and smell) while other organs (taste and touch) perceive stimuli by means of direct contact only.

Some sense organs can supplement other organs to a certain extent. For example, the acute sense of smell or touch can, to a certain extent, compensate poorly developed vision (eyes) and smell (nose).

Human sense organs.

Information received by the human brain from the sense organs forms the person's perception of the surrounding world and himself.

A human being receives information through six basic sense organs [8–10]:
1. Eyes (vision).
2. Ears (hearing).
3. Tongue (taste).
4. Nose (smell).
5. Skin (touch, sense of pain, temperature) [2].
6. Vestibular apparatus (sense of balance and position in space, acceleration, and sense of weight).

6 Receptors

Receptors are cells that process signals into nerve impulses. Receptor is a combination of dendrite terminals (nerve endings) of sensitive neurons, glia, specialized formations of intercellular substance, and specialized cells of other tissues, which in combination ensure transformation of the influence of external or internal environment factors (stimulant) into a nerve impulse.

The principle of receptors.

Types of stimuli for different receptors include light, mechanical deformation, chemicals, temperature changes, and changes in the electric and magnetic fields. In receptor cells (either nerve endings or specialized cells), the corresponding signal changes the conformation of sensitive molecules - cell receptors, which leads to change in the activity of membrane ion receptors and change in the membrane potential of the cell. If the nerve ending is the perceiving cell (the so-called primary receptors), then it usually leads to the membrane depolarization followed by generation of a nerve impulse. Specialized receptor cells of secondary receptors can both depolarize and hyperpolarize. In case of hyperpolarization, change in the membrane potential leads to decrease in secretion of the inhibitory transmitter influencing the nerve ending and, ultimately, to generation of the nerve impulse. In particular, such a mechanism is realized in the sensory elements of the retina.

Types of receptors.

There are several classifications of receptors.

According to the adequate stimulus:

- Chemoreceptors. They perceive the effects of dissolved or volatile chemical substances.
- Osmoreceptors. They perceive changes in the osmotic concentration of the fluid (usually, of the internal medium).
- Mechanoreceptors. They perceive mechanical stimuli (touch, pressure, stretching, vibrations of water or air, and others).
- Photoreceptors. They perceive visible and ultraviolet light.
- Thermoreceptors. They perceive cold or thermal stimuli.
- Electroceptors. They perceive changes in the electric field.
- Magnetic receptors. They perceive changes in the magnetic field.

The human being has the first six types of receptors. Taste and smell are based on chemoreception; touch, hearing, balance, and sense of the body position in the space are based on mechanoreception; vision is based on photoreception. Thermoreceptors are in the skin and some internal organs. Most interoreceptors start involuntary and in most cases unconscious and vegetative reflexes. Osmoreceptors are included in the regulation of kidney activity; chemoreceptors, which perceive pH and concentrations of carbon dioxide and oxygen in the blood, are included in the regulation of respiration, and so on.

7 Neuron

A neuron (from Greek νεῦρον - fiber, nerve) is a structural and functional unit of the nervous system. A neuron is an electrically excitable cell that processes, stores, and transmits information using electrical and chemical signals. A neuron has a complex structure and a narrow specialization. A cell contains a nucleus, a cell body, and processes (dendrites and axons). In the human brain, there are about 85-86 billion neurons. Neurons can connect with each other, thus, forming biological neural networks. There are receptor, effector and intercalary neurons.

8 Synapse

A synapse (Greek σύναψις, from συνάπτειν – connection, link) is a place of contact between two neurons or between a neuron and a signal-receiving effector cell. It serves for transmission of a nerve impulse between two cells; during the course of synaptic transmission, the amplitude and frequency of the signal can be regulated. Transmission of impulses is carried out chemically by means of mediators or electrically by ion passing from one cell to another.

The synapse creates discharges with a frequency of 1…100 Hz, the maximum value and exponentially decays, which is an alternating electric current generator. It generates an alternating magnetic field that changes the paramagnetic properties of gliacytes, which then provide associative memory by forming patterns in different isolated parts of the brain with pathways to the sense organs through the receptors.

9 Neuroglia

Glial cells are more numerous than neurons and make up at least half the volume of the central nervous system, but unlike neurons, they cannot generate action potentials. Neuroglial cells are different in structure and origin; they perform auxiliary functions in the nervous system providing support, trophic, secretory, delimiting, and protective functions. Glial cells have magnetic properties and contain paramagnetic oxygen-saturated solutions.

10 Cortex column

A cortex column (also called as a hypercolumn or cortical module) is a group of neurons located in the brain cortex perpendicular to its surface. Within one mini column, neurons have the same purpose, whereas a hypercolumn means the entire set of values for any set of parameters given by the receptors. A cortical module can denote a hypercolumn (V. Mountcastle) or a block of hypercolumns (D. Hubel and T. Wiesel).

The human neocortex.

The human neocortex contains six different layers, each of which can be identified by the type of neurons in it and their destination (inside the brain).

Modular organization of columns.

Modular organization of the columns was originally described by Vernon Mountcastle who suggested that the receptive fields of neurons located horizontally (relative to the cortex surface) at a distance of more than 0.5 mm (500 μm) from each other, do not overlap. Other experiments gave similar results (Buxhoeveden, 2002; Hubel, 1977; Leise, 1990, and others). Various estimates suggest that the hypercolumn is composed of 50 - 100 mini columns, each of which contains approximately 80 neurons.

11 Electromagnetic waves

The pulses generated by synapses and neurons are alternating electric current, which generates an alternating magnetic field.

The electromagnetic field in a vacuum is determined by the Maxwell equations:

$$\text{rot } \overline{E} = \frac{1}{c}\frac{\partial H}{\partial t}, \text{ div } \overline{H} = 0$$

$$\text{rot } \overline{H} = \frac{1}{c}\frac{\partial E}{\partial t}, \text{ div } \overline{E} = 0$$

This equation has solutions different from 0. Thus, electromagnetic field can exist even in conditions of absence of electric charges and in this case is called an electromagnetic wave.

For an electromagnetic wave, there is the d'Alembert equation (wave equation)

$$\Delta\overline{A} = \frac{1}{c^2}\frac{\partial^2 A}{\partial t^2}$$

or

$$\frac{1}{c}\frac{\partial\varphi}{\partial t} + \text{div } \overline{A} = 0$$

The equation has a relativistic invariant character: postulates that satisfy it in one reference frame satisfy it in all others.

Thus, temporally taken alternating electromagnetic field generated by the human brain by means of thought processes, existed in certain periods of the space-time continuum and was emitted into the surrounding space.

12 Field interaction

Because of finiteness of the velocity of interaction propagation within the force field, change in the position of one of the charged particles is reflected in the other only after a certain time interval.

Therefore, the force field reflects physical reality and has a measure value (time and force characteristics). Charged particles do not directly interact with each other.

Thus, the charged particle interacts with the field and subsequently, after a certain time interval, the field interacts with another charged particle.

In the case of motion of a unit charge (anion, cation), forces affect it or it affects the surrounding field by force.

$$E = -\frac{1}{C}\frac{\partial A}{\partial t} = -grad\varphi$$

where **A** is an operation work that must be performed over the charge in order to move it in space; **E** is the electric field strength.

H = rotA where, **H** is the intensity of the magnetic field.

We should note that **E** is a polar vector, and **H** is an axial vector.

Equation of a charge motion in an electromagnetic field (Lorentz force).

$$\frac{\partial p}{dt} = eE = \frac{l}{c}[VH]$$

where eE is a force of action of the electric field on the charge; it does not depend on the charge velocity and is oriented along the direction of the E field.

$\frac{e}{c} * [VH]$ is a force of the magnetic field on the charge; it is proportional to the velocity of the charge and is directed perpendicular to this velocity and to the direction of the magnetic field **H**.

Only the electric field acts on the charge produces; the magnetic field does not act on the charge moving in it because the force of action of the magnetic field on the particle is always perpendicular to its velocity. This property allows the magnetic field to exist always and not to expend its potential energy for acting on charged particles. Thus, change of the magnetic field forms associative memory and, simultaneously, is an apparatus for identifying and recognizing objects (as in neural chips or the coefficients of the trained neural networks (patterns).

13 Paramagnetic substances

Paramagnetic substances are substances that are magnetized in the external magnetic field in the direction of the internal magnetic field (**J↑↑H**) and have a positive magnetic susceptibility, but are much less than unity. Paramagnetic substances relate to strongly magnetic substances, their magnetic permeability significantly differs from unity. The paramagnetic substances have their own magnetic moments, which under the action of external fields are oriented along the field and thereby create a resultant field that exceeds the external field. Paramagnetic substances are drawn into a magnetic field. If there is no external magnetic field, paramagnetic substance is not magnetized, because atoms' magnetic moments are oriented quite randomly because of the thermal motion.

Paramagnetic substances include oxygen O2 (1.000017), air (1.000038), nitrogen oxide NO, manganese oxide and others. The authors believe that oxygen provides paramagnetic properties to glial cells; air with impurities of other gases is harmful to the brain. Paramagnetic resonance was discovered by E. Zavatsky in 1944, i.e. when introduced into a constant magnetic field of intensity **H**, a rotating electric charge receives an additional rotation about an axis parallel to the field with the frequency of

$$\Delta\omega = \frac{Pm}{L\alpha} * H$$

$\Delta\omega$ is a procession of the magnetic moment.

If different radio waves are passed through a paramagnetic body placed in a constant magnetic field **H**, then in conditions of coincidence of frequency of waves with the natural frequency, the resonance phenomenon will appear and intense energy absorption of the incoming wave will occur.

When recognizing an object, which is already stored in the memory, each memory region is broken up into cortex columns, where the neurons, axons, and the nerve endings of the dendrites are the conductor and the surrounding glial cells are paramagnetic solutions (saturated with oxygen). It is the memory that absorbs excessive energy during the neuron excitation and amplifies the current of the electrolytic reaction in its magnetic field. Therefore, because of the resonance, an acoustic signal is generated, which is perceived by the created local neural networks with the same natural frequency, i.e. the pattern is excited and it delivers the image recognized to the receptors of the 2nd signaling system; it means that we sense the object recognized once again.

14 Neural networks of the brain

1. Cortex columns - subnets as a repository of information.

The brain is divided into regions responsible for processing of visual, auditory, and other information and at the same time into cortex columns (subnets), which store information about every event.

Thus, subdivided neural networks are formed within each formation, the synaptic connections are much stronger and more numerous inside than between subdivisions.

A fully connected neural network containing N neurons is able to memorize $p=L_cN$ images, where L_c=1.185. Analysis of the stability of images stored in the neural network shows that if all images are orthogonal to each other, i.e. there is no mutual correlation between the images, then $p = N$.

Information about the general composite image of the object is stored in different regions of the brain.

2. Partial ordering of networks - chronological development over time.

Let us introduce a concept of eventfulness as chronological discrete existence and development of the human life, and, accordingly, formation of memory as the most acute and strong excitation of certain parts of the brain at certain periods of time.

If during the day events are recorded continuously, then the period of sleep is spent on ordering and extracting the most significant events from the records and fixing them in a dynamic long-term memory.

Formation and training of these neural networks occurs using a second signaling system under the control of a specific region of the brain.

We denote a set of subnets created this way and consisting of the patterns formed as S1 and a set of all networks representing all events in the human life as S. As time of pattern creation goes back and discrete memory can be built from continuous daytime wakefulness, then we can introduce a binary relationφ on the given set, which is called as partial ordering of this set (or a partial order).

Let us define the basic characteristics:

1. Individuality:

The relationship φ is a characteristic of particular individual because it is based on many memories or events S, which happened to this individual.

2. Generality:

The relationship φ has common properties inherent in all living beings. Let us denote φ by $\leq$.

$\forall x \in S$

P1. $x \leq x$ is reflexive for all $x \in S$.

P2. If $x \leq y$ and $y \leq x$, then $x = y$ is antisymmetric.

P3. If $x \leq y$ and $y \leq z$, then $x \leq z$ is *transitive*.

P1 means that if a given event occurred with a given person at a given moment and a pattern was created in the person's memory, then it could not participate at the same moment in a completely different event and in a completely different place.

P2 means that if an event occurred with a given person at a given specific time period and another event occurred with the same person at the same time period in another place, then it is the same event.

P3 is very important. If one event occurred with a given person before the next event came, and the third event occurred after the second one, it means that the first event occurred before the third one.

Thus, when building a discrete chronological chain of events in the memory of a particular person, as well as a chain of significant events for the whole community, we form the chronological memory

IOP Conf. Series: Materials Science and Engineering **363** (2018) 012022 doi:10.1088/1757-899X/363/1/012022

of a person in the form of patterns that connect multiple time and spatial scales related to the activity of the human brain.

Therefore, people can say, "It was before" or, "It was after".

A neural robot can exist precisely in a time scale, however, from the point of view of adaptation to the human society, it must also identify and record the most significant events for the surrounding people and use these phrases when communicating with people.

Let us consider some of the simplest properties of partially ordered sets and their projection on the mechanism of human memory formation.

a: The ratio inverse to the partial order $\varphi = \leq$ is again a partial order, which is called dual to the first one and is denoted as $\varphi^{-1} =$

Thus, by definition, $x \geq y$ if and only if $y \leq x$.

Partially ordered sets can be conveniently described by diagrams as shown in Figure 1.

Example:

P (3) is a set of all subsets of 3={ 1,2,3} partially relative to the ordered inclusion relation.

3 is today's state

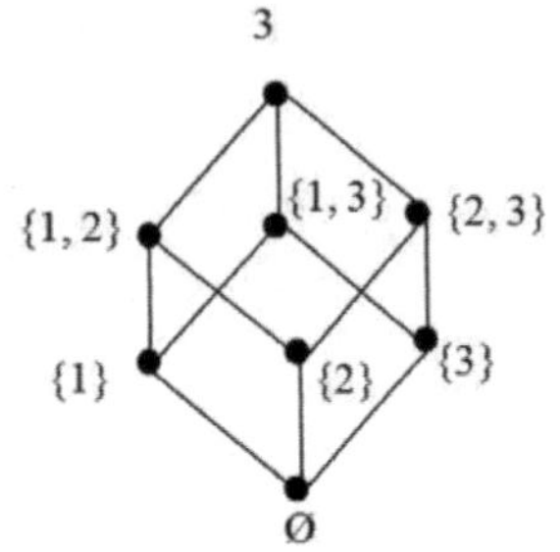

Figure 1. Diagram of a partially ordered set P{3}.

The duality principle (Theorem of theorems).

Theorem. In any general theorem about partially ordered sets of events, we can replace $\varphi=<$ by reverse $\varphi^{-1}=>$ without violating its truth.

If we consider the segments of the human life, we can assume that certain events will occur with him during the day or day-night period (month, year); their scheme is represented in the diagram. However, we can also introduce universal boundaries.

Elements *O* and *I* are called universal boundaries of a partially ordered set *S* (respectively upper and lower), if $0 \leq x$ and $x \leq 1$ for any $x \in S$.

O is the smallest, and *I* is the largest element of the set *S*.

Lemma. In any partially ordered set [S, ≤], no more than one minimal element and no more than one maximal element can exist.

Proving. Let *O* and *O** be the two minimal elements [S,≤]. Then 0≤ 0* and 0*≤ 0, according to P2; it means that O*= 0. It is analogous for *I*.

A person cannot live the life twice, or two different people cannot completely repeat the same event chronology, otherwise, this is the same person. According to this lemma, the event chronology can be very close for identical twins, because *O* and - *I* are very close in time and space.

Definition: Let P = [S,≤] be a partially ordered set, **a** and **b** be its elements. We say that **a** dominates over **b** if **a>b**, but there is no $x \in S$ so that **a> x> b.**

It means that if the event **a** follows immediately after the event **b**, then several different events *b* can occur (*b*2) at the next moment of time.

Theorem. Let a <b in a finite partially ordered set *P*. Then *P* contains at least one chain xo=a<x, <- <x<=b, in which each of the elements x_i (i = 1, ..., k) dominates on X_i-1.

IOP Conf. Series: Materials Science and Engineering 363(2018) 012022 doi:10.1088/1757-899X/363/1/012022

Thus, out of all the events that could happen to a given person, we can always construct an ordered chain of events in time and in space that had led this person from one point of the space-time continuum to a given point; moreover, no another event could be inserted into this chain of events.

By means of forming human memory out of the most significant events, we build an internal chronology of events in the form of a chain of patterns and using it, we can return to any point. It means the following.

Theorem. Let $[S,\leq]$, S $\{s,...s_i\}$ be a finite partially ordered set of events. Then the elements of *S* can be enumerated as $S=\{x, \quad x_n\}$, which implies that $i<j$ follows from $x_i<x_j$.

15 Determination of a pattern

The approach oriented on combining the principles of training and interpretability corresponds to the goals and objectives of Data Mining. The authors lay down the foundation of the modern technology of Data Mining (discovery-driven data mining) by the concept of patterns reflecting fragments of multidimensional relationships in data.

These patterns represent the regularity typical of the data subsamples, which can be compactly expressed in a human-readable form. Patterns are investigated using methods that are not limited to a priori assumptions about the sample structure and the form of distribution of the analyzed indicator values.

The experience can be accumulated in the form of formation, preservation, and correct application of patterns (images), which represent a library of the neural networks trained during the period of training or during a real life. For people, P (1) is important, i.e. the starting point formed in the childhood.

16 Application of patterns

Pattern-based thinking allows using almost the same program if it is implemented in the form of a neuroemulator or a neurochip and if the pattern coefficients files are being saved and loaded during the process of identification, recognition, etc.

In the artificial neural network *NeuroInformGeo*, the patterns were formed based on the already studied and drilled fields according to geophysical, geochemical, 2D, and 3D seismic data; new oil and gas fields were discovered in Eastern and Western Siberia and the Sakha (Yakutia) Republic (Russia) [5].

In the artificial neural network *NeuroCiber*, the patterns were formed based on the phonemes of a specific language and specific people, as well as on photos taken by the robot during the acquaintance; combination of audio-video based recognition led to a sharp improvement of identification and recognition of subjects as shown in Figure 2 [6 ,7.10].

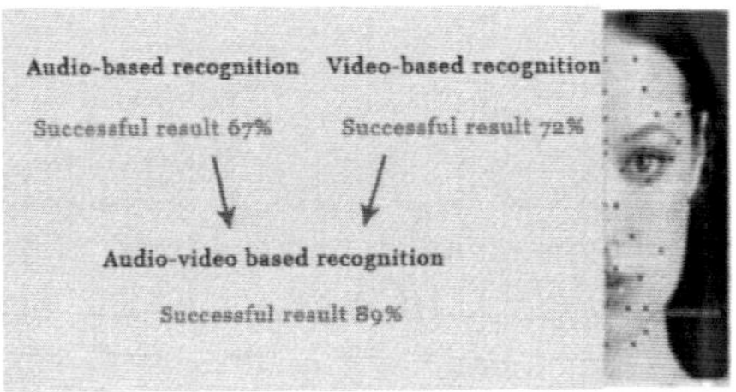

Figure 2. Audio-video based recognition.

In the artificial neural network *NeuroFinExpert*, the patterns were formed based on financial stability of groups of enterprises; it allowed analyzing and predicting economic, social, economic, and political results over time.

The latest SyNAPSE chip, introduced on August 7, 2014, has sensory capabilities at incredibly low energy level. IBM built a new chip with a brain-like computer architecture, which actuates 1 million neurons and 256 million synapses. It is the largest chip IBM has ever built at 5.4 billion transistors, and has an on-chip network of 4,096 neurosynaptic cores. Yet, it only consumes 70mW during real-time operation — orders of magnitude less energy than traditional chips.

17 Conclusion

The brain sings. By exchanging musical phrases between different regions of the brain responsible for processing information from various sense organs, it synchronizes the composite image (pattern) and, when new information passes through it, the answer comes based on the previous accumulated experience.

There is a number of interesting points arising from this model of brain functioning. Sigmund Freud was right seeking the event in the depth of consciousness, that is, the pattern associated with this event and affecting the current psychological behavior of a patient.

IBM specialists draw an interesting analogy between conventional computer systems and the left hemisphere of the human brain. Neuromorphic systems can find individual patterns in large data sets and interpret this information as the right hemisphere of the brain. In the near future, scientists will be able to combine the capabilities of traditional computers and neuromorph chips and to create a single super effective structure – an artificial intelligence [4].

Acknowledgements

The research presented in this paper was supported by the Russian Foundation for Basic Research (grant No 16-29-04388) and by "The Tomsk State University competitiveness improvement programme" grant (No № 8.2.24.2018)

The authors are grateful to Tatiana B. Rumyantseva from Tomsk State University for her assistance in preparing the paper.

References

[1] Gafurov O M, Gafurov D O, Gafurov A O, Pankov M V, Bitner A K, Krasilnikova N B 2012 Patent of the Russian Federation for invention No. 2477499 "A Method for Determining Location of Production Wells During the Development of Hydrocarbon Fields", application No. 20111125437

[2] Gafurov O M, Gafurov D O, Gafurov A O 2011 Certificate of the Federal Service for Intellectual Property for the state registration of the computer program "The Neural Network-Based Program for Processing Seismic Acoustic and Geophysical Data in .dat Grids Format Used for Formation of Etalons and Calculation of Patterns", No. 2011613919

[3] Gafurov O M, Gafurov D O, Gafurov A O 2011 Certificate of the Federal Service for Intellectual Property for the state registration of the database: "The Database for the Block of Automated Formation of Etalons in the Intellectual Geoinformation System "NeuroInformGeo" No 201162031

[4] Gafurov O, Gafurov D 2016 *MATEC Web of Conferences* 01086

[5] Gafurov O, Gafurov D 2016 *MATEC Web of Conferences* **79** 01055

[6] Gafurov A O 2004 *The 58th Republican scientific conference of young scientists and graduate students dedicated to the 70th anniversary of the Kazakh National University* 122

[7] Gafurov O M, Gafurov A O 2004 *All-Russia scientific and technical conference "Science. Defense. Industry"* 34

[8] Cox I J, Ghosn J and Yianilos P N 1995 *Feature-based face recognition using mixture distance*, NEC Research Institute, Technical Report № 95-09, 58

[9] Kuharev G A 2001 *Biometric Systems: Methods and Tools for Identification of a Human Person* (SPb., Polytechnics)

[10] Gafurov O M, Syryamkin V I, Gafurov A O, Stolyarova S S 2012 *Telecommunications and radio engineering* **71** 1565–1574

Application of the magnetic fluid as a detector for changing the magnetic field

D Zyatkov[1,2], A Yurchenko[1,2], V Yurchenko[2,3], V Balashov[2,3]
[1]National Research Tomsk Polytechnic University, Tomsk, Russia
[2]National Research Tomsk State University, Tomsk, Russia
[3]Research Institute of Semiconductor Devices Tomsk, Russia
E-mail: zyatkov.88@mail.ru

Abstract. In article the possibility of use of magnetic fluid as a sensitive element for fixing of change of induction of magnetic field in space is considered. Importance of solvable tasks is connected with search of the perspective magnetic substances susceptible to weak magnetic field. The results of a study of the capacitive method for fixing the change in the magnetic field on the basis of a ferromagnetic liquid are presented. The formation of chain structures in the ferrofluid from magnetic particles under the influence of the applied magnetic field leads to a change in the capacitance of the plate condenser. This task has important practical value for development of a magnetosensitive sensor of change of magnetic field.

1. Introduction

Magnetic fluids are artificially formed and specially structured media with unique electrical and magnetic properties. Such properties should be understood as special values of the physical parameters of the medium: dielectric ε and magnetic μ permeabilities, spatial structurization, depending on the size and shape of magnetic and dielectric particles, the possibility of controlling the parameters of the medium as a result of external influences. The high mobility of magnetic particles in the fluid ensures their strong sensitivity to low magnetic fields and gives a significant advantage in comparison with solid-state analogs [1, 2].

Now the increased interest in the magnetic and magnetodielectric fluids (MF) shown from theorists and experimenters is caused by the fact that these substances have properties which are in many respects unique and difficult predicted. The uniqueness of these properties and the possibility of practical use in various industries contribute to the development of fundamental research on their study and practical application in complex technical problems. For example, using MF developed technologies for cleaning water surfaces from oil products in case of accidents and catastrophes; MF allowed to create dampfers and shock absorbers with controlled rigidity; MF became the basis for creating acceleration, tilt, level sensors; on the basis of the magnetic fluid, magnetic-fluid bearings, suspensions and bearings have been created; at present, physicists from the Michigan University of Technology have developed a new type of engine for small space vehicles. They resemble ionic engines, but the working substance in them is an ionic magnetic fluid. Thus, the use of a magnetic fluid expands, there are new areas of their application [3–8].

Usually, magnetic fluids investigate in strong magnetic fields ($B = 10^{-3} - 10^{3}$ T). The study of structuring mechanism of magnetic particles in the fluid matrix under the influence of a weak magnetic field ($B = 10^{-6} - 10^{-9}$ T) and creation of model of omposition magnetic substance for a sensitive element of the capacitive sensor of magnetic fields have a considerable scientific interest.

The importance of current task is connected with search of the perspective magnetic substances susceptible to weak magnetic field [9,10,12].

2. Object of experimental studies

In experimental studies the magnetic fluid on the basis of a polimetilfenilsiloksan (PFMS-4) containing particles of carbonyl iron dispersion of 3 microns was applied. Concentration of particles of iron in PFMS-4 doesn't exceed 10% and 40%. Preparation of magnetic fluid was carried out by mechanical and ultrasonic agitation.

The next step was to located suspension on a ceramic plate with a structure of condenser. The condenser is made of gold. A photo of the capacitor is shown in Figure 1.

Figure 1. Capacitor on a ceramic plate.

The scheme of the capacitor is shown in Figure 2.

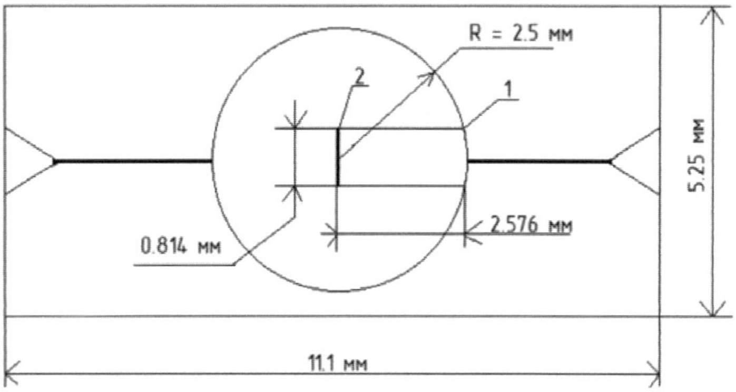

note
1. d = 0,007 мм
2. d = 0,076 мм

Figure 2. Scheme of the condenser.

The thickness of the a ceramic plate is equal to 470 microns. The thickness of gold plating is equal to 6.21 μm.

Solid particles in a magnetic fluid experience the action of gravity and Archimedes force. Gravity in the case of spherical particles is equal to:

$$F_g = -\frac{4}{3}\pi r^3 \rho_p g$$

Archimedes force in case of spherical particles is equal:

$$F_{Arch} = \frac{4}{3}\pi r^3 \rho_f g$$

where r – particle radius, ρ_p and ρ_f – density of a particle and fluid respectively.

Under the action of sedimentation force, the particles acquire velocity. This velocity leads to the appearance of a frictional force which is directed opposite to the velocity of the particles in the fluid medium. For spherical particles friction force can be calculated according to the Stokes law:

$$F_{fr} = -6\pi r\eta v$$

where η – the dynamic viscosity of the fluid, v – the velocity of the particle.
Thus, the total force acting on particles in a magnetic fluid is equal to:

$$F = -\frac{4}{3}\pi r^3 \rho_p + \frac{4}{3}\pi r^3 \rho_f g - 6\pi r \eta v$$

The equation of motion (settling) of particles along the z axis is written as:

$$\begin{cases} M\ddot{z} = -\frac{4}{3}\pi r^3 \rho_p g + \frac{4}{3}\pi r^3 \rho_f g - 6\pi r \eta \dot{z}; \\ z(0) = z_0; \\ \left.\frac{dz}{dt}\right|_{t=0} = v_0. \end{cases}$$

Solving this equation, we obtain:

$$z(t) = z_0 + \left(v_0 - \frac{\alpha}{\beta}\right)t + \frac{\alpha}{\beta^2}\left(1 - e^{-\beta t}\right);$$
$$v(t) = v_0 - \frac{\alpha}{\beta}\left(1 - e^{-\beta t}\right). \qquad (1)$$

where $\alpha = g(1 - {\rho_ж}/{\rho_ч})$, $\beta = {6\eta\pi r}/{M}$

A magnetic fluid consisting of carbonyl iron particles and PFMS-4 was placed in a cylindrical vessel 0.5 cm high. Figures 3 and 4 show the graphs $z(t)$ and $v(t)$ constructed according to formula 1 for a magnetic fluid in a cylindrical vessel. The radius of the carbonyl iron particle is equal to 2 μm, the density of the PFMS -4 is equal to 1.1 g/cm^3, the density of the particles is equal to 7.5 g/cm^3, the dynamic viscosity of the fluid is equal to 6.6 g/(s * cm).

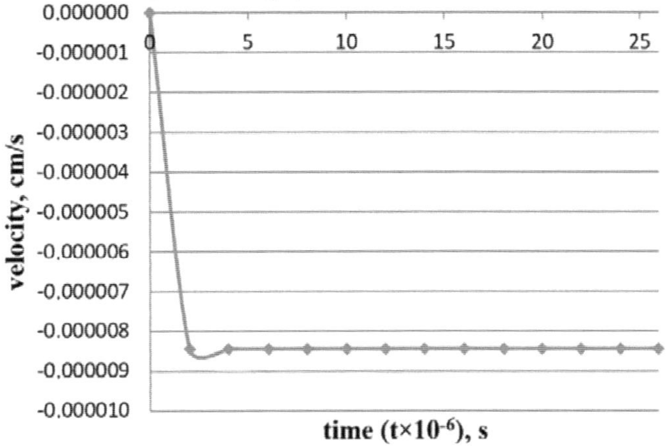

Figure 3. Graph of the dependence of the velocity of motion of particles on time.

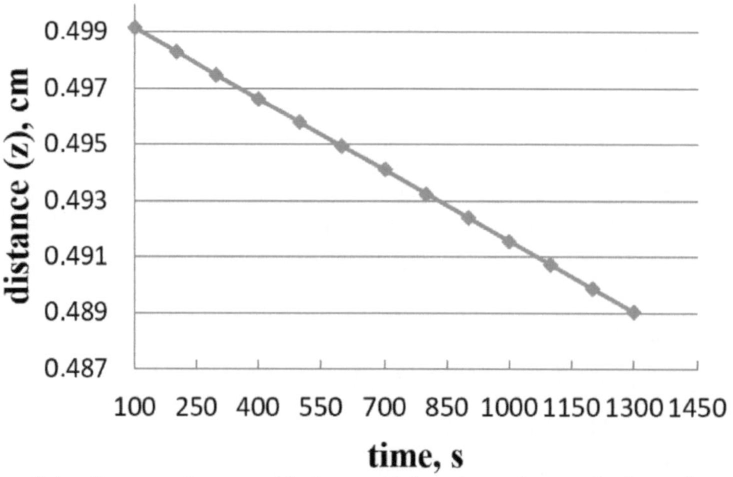

Figure 4. The plot of the distance the passable by particles along the z axis from time.

At the initial moment of time, because of the low settling velocity, the resistance of the fluid is small, and the particles move at an accelerated rate. With further increase in velocity, the frictional force becomes equal to the force of sedimentation. It can be seen from graph 3 that the particle velocity during a time of the order of 10-6 s is established and move with a constant speed. It can be seen from graph 4 it is visible that the position of the particle along the z axis changes by 0.001 cm in 100 s. It should be noted that in the preparation of a magnetic fluid, a stabilizing agent (surfactant) was not used. The absence of surfactants leads to the sedimentation of particles and the stratification of the magnetic fluid over time. Several minutes were spent for performance of measurements. It allows to consider magnetic liquid as steady system and to neglect influence of sedimentation of particles in the course of measurement of change of magnetic field.

3. Method and result of experimental study

Each atom of magnetic substance produces a tiny magnet or magnetic dipole, so total magnetization of substance is achieved when all the single atomic magnets are built into a certain order. In this way, the presence of the magnetic moments leads to their orientation longer axis along direction of field. Due to their interaction, they can uniting be combined chain aggregates under effect of the magnetic field (Figure 5) [11].

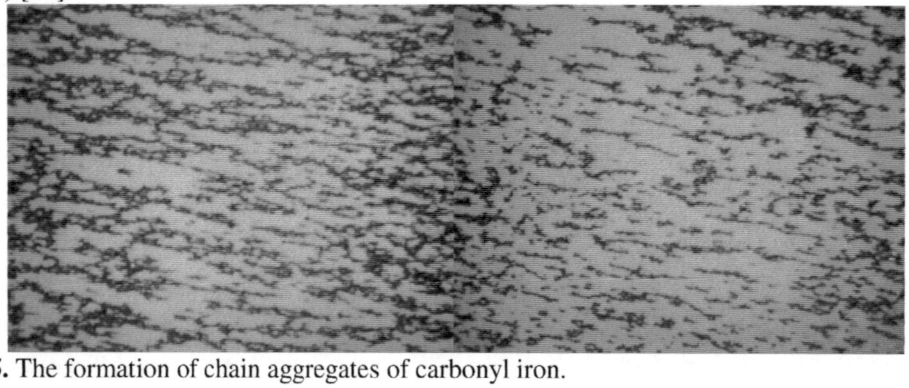

Figure 5. The formation of chain aggregates of carbonyl iron.

It is assumed that the formation of chain structures can lead to significant changes in the magnetic and other properties of ferrofluids which influence on change of capacity of the condenser filled with magnetic liquid. The capacitance measurement was carried out on an E7-12 device. To study the effect of a magnetic fluid on the electrical parameters of a cell, it was subjected to an external magnetic field. Figure 6 shows the scheme of the experiment.

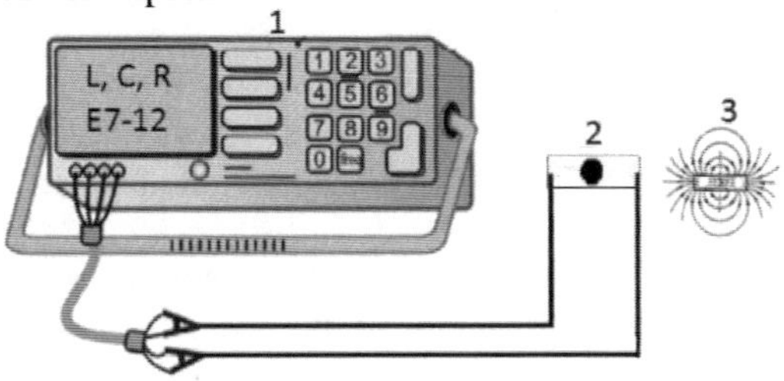

Figure 6. Experimental setup for studying the effect of a magnetic field on a magnetic fluid in a capacitor, where 1 is a meter of L, C, R type E7-12; 2 – a condenser on a ceramic plate with a magnetic fluid; 3– the magnet.

The magnetic field was created by a permanent magnet. To determine the strength of the magnetic field (magnetic induction) of the magnet acting on the measuring cell, a microteslameter MT-10 was used.

The coordinates of the position of the magnet are (x: y: z): (i: 0: 0); (-i: 0: 0); (0: i: 0); (0: -i: 0); (0: 0: i), where i = 20, 15, 10, 5 mm (the distance from the magnet to the sensor). The distance was measured on millimeter paper.

Before the experiment by influence of the magnet field on the electrical parameters of the cell with the magnetic fluid, we have measured induction of magnetic field in the place of an experiment. The magnetic field induction was 84 μT. Then a cell with a magnetic fluid of carbonyl iron was placed at the site of the experiment. The magnetic field of the magnet was applied to the measuring cell in five directions ((i: 0: 0); (-i: 0: 0); (0: i: 0); (0: -i: 0); (0: 0: i)). In Table 1 values of relative change of capacity the condenser with magnetic fluid depending on concentration of particles in magnetic fluid and induction of magnetic field are given.

Table 1. Relative change of capacity from the position of the magnet in space

		The concentration of particles in a magnetic fluid, 10 %				The concentration of particles in a magnetic fluid, 40 %		
Capacitance of a capacitor with a magnetic fluid without the action of a magnetic field, pF		0.993				1.118		
Distance from magnet to sensor (mm)		20	15	10	5	20	10	5
The magnetic field strength of a magnet, (μT)		35.9	50	65.5	105	35.9	65.5	105
Location of the magnet relative to the condenser with magnetic fluid	(i:0:0)	0.993	0.993	0.995	0.998	1.118	1.150	1.15
	(-i:0:0)	0.993	0.993	0.996	1.005	1.118	1.135	1.142
	(0:i:0)	0.993	0.993	0.996	1.001	1.118	1.150	1.170
	(0:-i:0)	0.993	0.994	0.999	1.002	1.118	1.135	1.167
	(0:0:i)	0.993	0.994	1.001	1.008	1.118	1.138	1.180
The relative change in capacitance (ΔC.%) under the influence of the magnetic field of the magnet		0	0.04	0.44	0.98	0	2.1	3.9

4. Conclusion

Thus, the results of the studies showed the possibility of detecting the change in the magnetic field produced by the magnet. The change in capacitance of a capacitor with a magnetic fluid is caused by the structurization processes proceeding in them. Table 1 shows that the sensor element based on the capacitor plate reacts to the magnetic field in all directions. In this case, the orientation sensitivity of the sensor element is practically independent of the position of the magnet in space. However It should be noted that there is a shortcoming, namely after influence by magnetic field on solution of a ferromagnetic, iron particles adopt the chaotic provision not at once. Because of this, further measurement accuracy suffers. For increase of sensitivity to weak magnetic field it is necessary to use particles of an anisotropic form with high magnetic permeability [9,13]. In order that energy of an attraction between particles didn't exceed energy of the chaotic movement, it is necessary to add to ferromagnetic liquid the substances consisting of polar organic molecules which create on a surface of disperse particles of the adsorption-solvate layers. This will also preserve the stability of the magnetic fluid and the particles will not settle in the gravity field, so that the magnetic fluid can retain its properties for a long time.

Acknowledgments

Sections 1 and 2 of the article are fulfilled by D Zyatkov, V Yurchenko and V Balashov, and section 3 - A Mehtiyev and A Yurchenko.

The work of D Zyatkov, V Yurchenko and V Balashov is supported by the Russian Science Foundation under grant (project No. 18-19-00268).

This study (research grant № 8.2.24.2018) was supported by the Tomsk State University competitiveness improvement programme.

References

[1] Taketomi S and Chikazumi S (1988) *Magnetic Fluids-principle and Application* (Tokio: Nokkan Kogyo Shinbun)

[2] Liao W H, Krueger D A 1979 *Journal of Colloid and Interface Science* **70**(3) 564–576

[3] Zyatkov D, Yurchenko A, Balashov V, Yurchenko B and Borisov A 2017 *Progress In Electromagnetics Research Symposium* 2707–2711. doi: 10.1109/PIERS.2017.8262211

[4] Kontarev A V 2012 *Progress of the modern natural sciences* (10) 67–70

[5] Jacksona B A *et al* 2017 *Physics of Fluids* **29** 064105 doi: 10.1063/1.4985141

[6] Bobrovitskiy D A, Demenkova L G 2014 *Modern state and problems of natural sciences: collection of works of the All-Russian scientific-practical conference of young scientists, postgraduate students and students* 203–205

[7] Senatskaya I I 2012 *Chemistry and life* (10) 43–47

[8] Scherer C, Figueiredo Neto A M 2005 *Brazilian Journal of Physics* **35(3A)** 718–727

[9] Zyatkov D *et al* 2017 *J. Phys.: Conf. Ser.* **881** 012037

[10] Zyatkov D, Yurchenko A and Yurchenko E 2017 *Progress In Electromagnetics Research Symposium* 3176–3181 doi: 10.1109/PIERS.2017.8262304

[11] Zubarev A Yu, Iskakov L Yu 1995 JETP **80 (5)** 857–866

[12] Yurchenko A *et al* 2016 *MATEC Web Conf.* **79** 01085

[13] Mekhtiyev A *et al* 2016 METALURGIJA 55 47–50

IOP Conf. Series: Materials Science and Engineering **363** (2018) 012024 doi:10.1088/1757-899X/363/1/012024

Ultrasound sounding in air by fast-moving receiver

D Sukhanov, N Erzakova
Radiophysics Faculty, Tomsk State University, Tomsk, Russia
E-mail: sdy@mail.tsu.ru

Abstract. A method of ultrasound imaging in the air for a fast receiver. The case, when the speed of movement of the receiver can not be neglected with respect to the speed of sound. In this case, the Doppler effect is significant, making it difficult for matched filtering of the backscattered signal. The proposed method does not use a continuous repetitive noise-sounding signal. generalized approach applies spatial matched filtering in the time domain to recover the ultrasonic tomographic images.

1 Introduction

Ultrasonic sounding systems in the air are widely used in ultrasonic machine vision tasks for the robot orientation in space at different speeds [1]. Modern cars robots rely on lidar [1, 2] and radars [3, 4] in the orientation in space, but this solution is costly. Ultrasound systems are used only at low speeds [5–7]. In addition, the range of ultrasound systems is not large, making them difficult to use at high speeds. Limiting speed with ultrasonic sensing due to the small speed of sound. During the period until the signal returned from the target locator time to shift in space at a distance of more than a quarter of the wavelength [8, 9]. But for unambiguous images by monostatic sounding measurements necessary steps less than a quarter wavelength. If we apply a short pulse signal for sounding, it must be repeated each time the system is shifted by a quarter wavelength. Since the motion is fast repetition period is reduced so that the spatial distance in distance between adjacent pulses is much smaller than the depth sensing. This creates a ghosting of images of objects in range and does not allow one to visualize the object. The solution to this problem by increasing the pulse repetition rate leads to a sparse synthetic aperture, and consequently to an increase in the level of artifacts. One way of overcoming this problem would be the use of broadband probing signals of different shapes. Such signals allow to distinguish between the different positions of the probing system. For example, use of wideband signals with different carrier so that the signal spectra do not overlap. However, this method does not provide an ultra-wideband sensing and obtain the best resolution in range.

It is proposed to develop methods of ultrasonic sensing to visualize the scattering inhomogeneities in fast motion radar system on the basis of noise-like signals to enhance the uniqueness of the measurement. Since this signal is a UWB obtain the maximum possible range resolution. Furthermore, the signal is non-repetitive, increasing visualization unambiguous in range.

2 System overview

It is proposed to consider the scheme of measurements shown in Figure 1. In the scheme of measurement used linear scanner which moves the receiver with the speed V. In this case, the emitter is not mobile and is in the center of the range of movement of the receiver. The test facility is located in the *h* range.

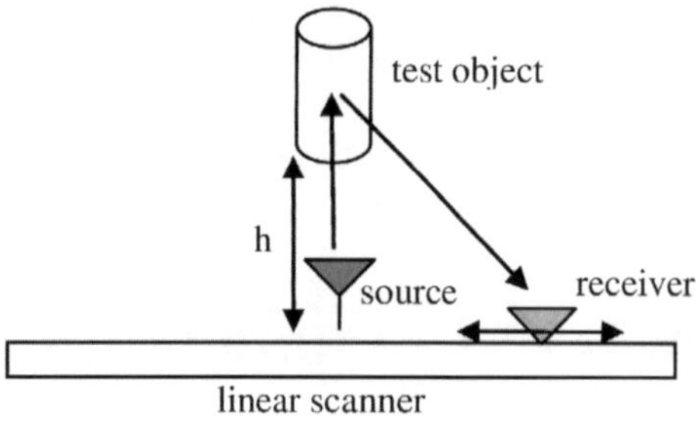

Figure 1. Measurement scheme.

We assume that the transmitter and receiver are not isotropic directed and describes the free space Green's function. Consider the problem in the approximation of single scattering. Let the signal emitted by the function $S(t)$. We represent the distribution of the scattering inhomogeneities as a function $p(x, y)$. The signal $U(t)$ at the receiver in the approximation of single scattering can be written as:

$$U(t) = \iint p(x, y) \frac{S\left(t - \frac{1}{c}\left(|\mathbf{r}_S - \mathbf{r}| + |\mathbf{r}_R - \mathbf{r}|\right)\right)}{|\mathbf{r}_S - \mathbf{r}| \cdot |\mathbf{r}_R - \mathbf{r}|} dxdy \, , \quad (1)$$

where c – the speed of sound in the environment; $\mathbf{r}_S = (x_s, b)$ – coordinates of the emitter; b – distance of the radiator from the scan line; $\mathbf{r} = (x, y)$ – the coordinates of the integration in the environment; $\mathbf{r}_R = (Vt, 0)$ – receiver coordinates.

It was numerically modeled noise-sounding signal with a frequency band of 37 to 43 kHz. The waveform is represented in Figures 2 and 3.

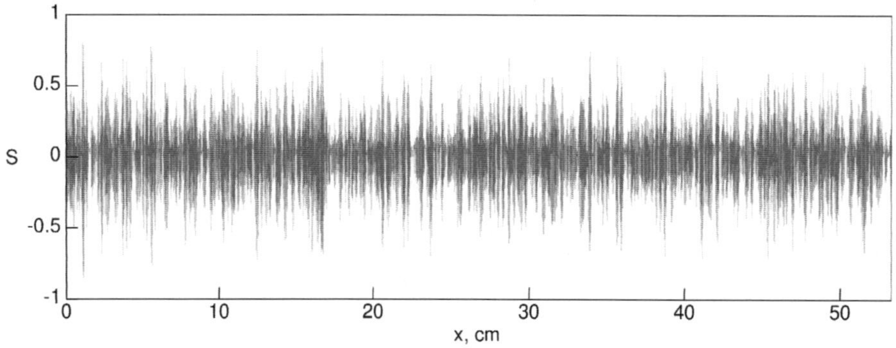

Figure 2. Modeled Noise-like sounding signal.

On the basis of the direct problem solution by the equation (1) was modeled receiver signal in the presence of two scattering objects at a distance of 50 cm from the scan line. During the scanning receiver has moved a distance of 90 cm at a speed of 17 m/s. The speed of sound was taken to be 330 m/s.

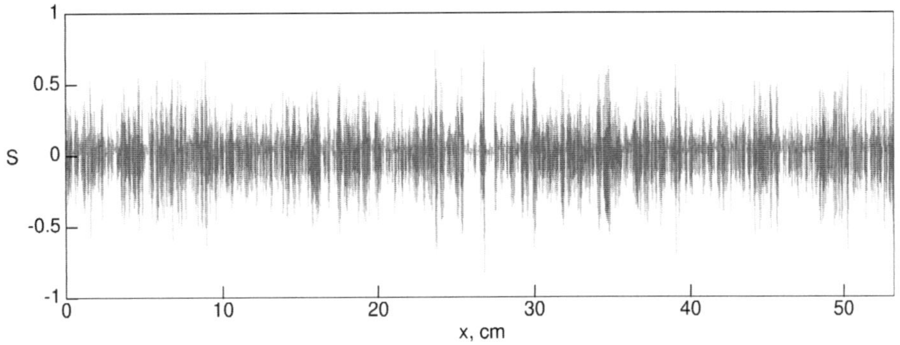

Figure 3. Modeled Noise-like signal in receiver.

It can be observed that the signal is always present, and it will not be repeated, while maintaining a broad band, and therefore has a lot of energy, which gives reason to expect a high level of signal-to-noise ratio.

3 Solving the inverse problem

Under the inverse problem solution refers to the determination by the measured signal $U(t)$ functions $p(x,y)$. To solve this problem has a method of spatial matched filtering in the time domain. According to this method, the solution can be written as:

$$p(x,y) = \int_{t_{\min}}^{t_{\max}} U(t) S\left(t - \frac{1}{c}\left(\left|\mathbf{r}_S - \mathbf{r}\right| + \left|\mathbf{r}_R - \mathbf{r}\right|\right)\right) dt\,, \qquad (1)$$

where $t_{\min}$ – the start time of the measurement; $t_{\max}$ – end time of measurement.

According to this equation, it was restored image of two lenses at a distance of 50 cm, which is shown in Figure 4.

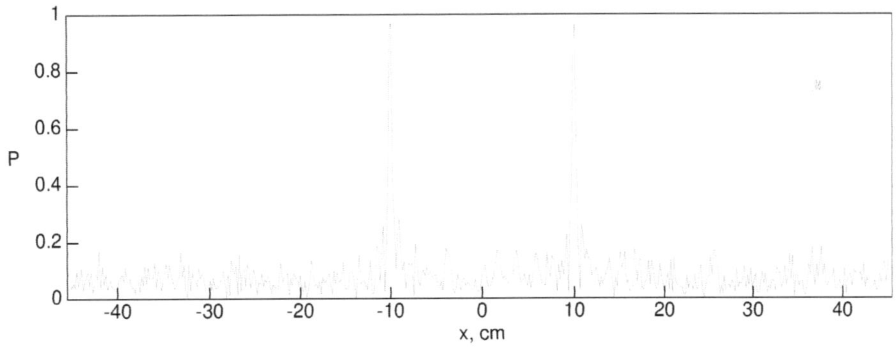

Figure 4. Reconstructed image of two test objects.

In Figure 4 you can see that the resolution obtained in the scanning plane of the order of 1 cm which is close to the diffraction limit for the considered frequency range. Of course, the presence of artifacts associated with the use of noise-like signal.

To estimate range resolution numerical simulation was carried out with two point lenses at ranges of 40 and 50 cm. Note that Spreader located opposite the center of the finish. Figure 5 shows the reconstructed image of two lenses at different distances.

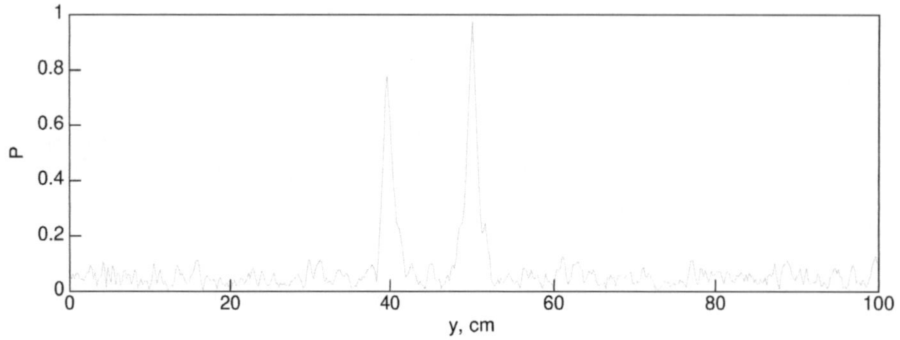

Figure 5. Reconstructed image of two test objects on distance.

According maxima width can estimate range resolution of 5 cm, which corresponds to the expectations for bandwidth.

Numerical simulation results show that the proposed method provides a resolution of the images is close to the theoretical limit. Also, the receiver allows to consider the movement speed at an arbitrary speed of sound.

To test the ability to visualize images of extended objects were numerically simulated probe at a speed of 4 m / s in the band between 20 and 26 kHz. The object length of 40 cm was placed at a distance of 50 cm from the scanning system parallel to the axis of motion. After processing according to the equation (1) was obtained image of the object represented in Figure 6.

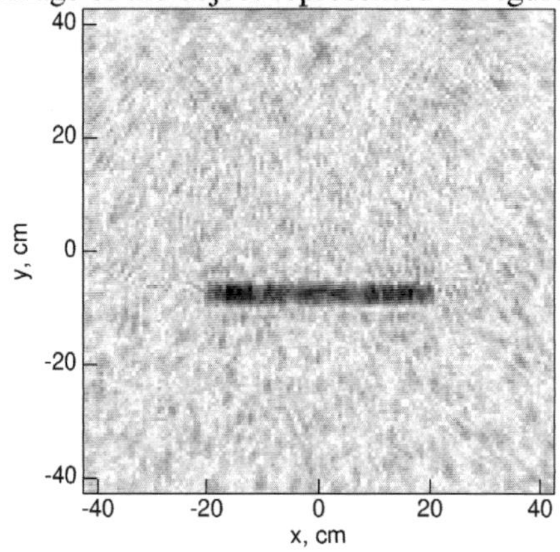

Figure 6. Reconstructed image of horizontal plane object.

It should be noted that the reduced signal to noise ratio by increasing the speed of the receiver. This is due to the fact that the duration of the signal is reduced with respect to the extent of the restored image.

4 Conclusion

The proposed method is an ultrasound probe in the air when moving at high speed receiver using noise-like probing signals. Image restoration method based on matched filtering. Numerical simulation results showed the possibility of obtaining the spatial resolution is close to the theoretical limit. By increasing the speed of movement of the receiver level artifacts in images produced increases.

Acknowledgment

The research is supported by Tomsk State University.

References

[1] Li J, Bao H, Han X, Pan F, Pan W, Zhang F, Wang D 2016 *Multimedia Tools and Applications*, **75(1)** 1–23 doi: 10.1007/s11042-014-2221-x
[2] Barfoot T D, McManus C *et al* 2016 *Springer Tracts in Advanced Robotics* **114** 487–504 doi: 10.1007/978-3-319-28872-7_28
[3] Schuster F *et al* 2016 *IEEE Intelligent Vehicles Symposium Proceedings* 7535485 doi: 10.1109/IVS.2016.7535485
[4] Zhu Y 2015 *16th International Workshop on Mobile Computing Systems and Applications* 75-80 doi: 10.1145/2699343.2699363
[5] Jong-Hann Jean *et al* 2013 *IEEE International Conference on Mechatronics and Automation* 1234 – 1238 doi: 10.1109/ICMA.2013.6618090
[6] Xuming Pei *et al* 2010 *IEEE International Conference on Mechatronics and Automation* 1245 – 1249 doi: 10.1109/ICMA.2010.5589942
[7] Bukhari Ilias *et al* 2016 IEEE Symposium on Computer Applications & Industrial Electronics (ISCAIE) 189 – 194 doi: 10.1109/ISCAIE.2016.7575061
[8] Aiordachioaie D *et al* 2014 *Autonomous Vehicles: Intelligent Transport Systems and Smart Technologie* 57–79
[9] Juha Ylitalo *et al* 1995 *Ultrasonics* **34**(2–5) 331–333

IOP Conf. Series: Materials Science and Engineering **363** (2018) 012025 doi:10.1088/1757-899X/363/1/012025

Main principles of developing exploitation models of semiconductor devices

A V Gradoboev and A V Simonova
National Research Tomsk Polytechnic University, Tomsk, Russia
E-mail: gradoboev1@mail.ru

Abstract. The paper represents primary tasks, solutions of which allow to develop the exploitation modes of semiconductor devices taking into account complex and combined influence of ionizing irradiation and operation factors. The structure of the exploitation model of the semiconductor device is presented, which is based on radiation and reliability models. Furthermore, it was shown that the exploitation model should take into account complex and combine influence of various ionizing irradiation types and operation factors. The algorithm of developing the exploitation model of the semiconductor devices is proposed. The possibility of creating the radiation model of Schottky barrier diode, Schottky field-effect transistor and Gunn diode is shown based on the available experimental data. The basic exploitation model of IR-LEDs based upon double AlGaAs heterostructures is represented. The practical application of the exploitation models will allow to output the electronic products with guaranteed operational properties.

1. Introduction

Robotic centers, or more precisely control systems of the robotic centers, are broadly used at nuclear power plants, on different space vehicles etc. That is, they used in the instances that do not allow a working man to provide the define work. In addition, the control systems of the robotic centers are exposed to impact of combined and complex influence of various types of ionizing irradiation and long-term operation factors (electrical fields, temperature gradient, mechanical stress etc.) [1–7]. In this case, we interpret that complex influence is the simultaneous impact of two or more radiation factors. Accordingly, the combined influence is the impact of two or more radiation factors spread out over a period of time [8]. We note, that interest in combined influence of ionizing irradiation on various semiconductor devices increases [9–16].

Nowadays, there is practically absent the possibility to guarantee the useful quality of the control systems of the robotic centers under the complex and combined influence of damaging factors.

The reliability and radiation resistance problems take on particular importance to ensure the guaranteed security of nuclear power facilities, radiation technology objects, repositories for radioactive substance etc.

Moreover, the electronic products (EP) are the main components for control systems of various robotic centers. Therefore, the performance characteristics of the EP can be represented as a set of following main properties: set of parameters and characteristics; radiation resistance; reliability (failure-free operation, durability and storability).

Therefore, the operation of the EP under these conditions can be characterized by exploitation model (EM) which should include: radiation model (RadM) and reliability model (RelM).

RadM (RelM) of the EP is a system of theoretical, empirical or/and semi-empirical laws, outlining correlations of basic material characteristics and/or parameters and characteristics of the EP with radiation resistance (reliability) indexes.

Before working out the EM of the EP it is necessary to develop the EM of basic components of the EP, which are discrete semiconductor devices. A list of basic discrete semiconductor devices used in the various EP is sufficiently small. In particular, this list can be limited to the following devices: Gunn diode; Schottky barrier diode; p-n-junction diode; light-emitting diode (LED); photodiode; Schottky field-effect transistor; p-n-p (n-p-n)-junction transistor and Hall-effect transducer.

Furthermore, the list of used semiconductor materials is also small enough. In addition, silicon and gallium arsenide are used as basic semiconductor materials. We note, that the heterostructures are abundantly used recently, for instance, AlGaAs, AlGaInP, InGaN, AlGaN [17].

The main methods of manufacture of device structures are vapor phase, liquid-phase and molecular beam epitaxy, ion implantation and various diffusion technology. As is evident, there are also few methods of devices structures formation, because all the rest are particular causes of listed methods.

Therefore, the purpose of this work is to develop the main principles of exploitation models of semiconductor devices (EM-SD), which allow to predict the change of criterion parameters under the influence of ionizing irradiation and factors of long-term operation with their combine and complex influence. Furthermore, the purpose of the work is to present the practical realization of the basic principles through the exploitation model of LED (EM-LED).

In this case, criterion parameter of the device is a parameter which determines its durability, at which point, it is the most sensitive to the influence.

2. Exploitation model of semiconductor devices

This section describes the basic principles of EM-SD development, which can be define in the following way:

- based on the theory;
- based on the semi-empirical laws, which establish relation between parameters of basic materials (parameters and characteristics of the devices) and radiation resistance (reliability) indexes;
- based on the experimental laws, which describe the change of criterion parameters of the device under the influence of ionizing irradiation (during long-term operation).

We note that it is required to ensure the following problem solving in the making of EM-SD:

- to determine the requirements to parameters and characteristics of basic semiconductor material;
- to establish the criterion parameters and characteristics of the device and basic device structures, which define radiation resistance (reliability) of the device;
- to detect the requirements to parameters and operational conditions used in the manufacture of the devices, which can influence over rate of defect introduction under influence of ionizing irradiation and during long-term operation factors;
- to take into account the influence of design factors on radiation resistance and reliability of the devices.

At present time there is sufficiently large number of computer models, which allow to calculate the formation of various radiation defects in condensed matter [18–21]. The obtained results correlate with experimental research results. However, the discrepancies do not allow to use these models as a basis for development of the RadM.

The RadM can be create on the basis of semi-empirical laws established relations between changes in the criterion parameters of the device under ionizing irradiation and electrophysical with geometrical characteristics of the device structures. The RadM allows to get the quantitative estimation of radiation resistance of the device in establishing the sufficient correction relations.

Furthermore, the RadM can be developed from experimental laws outlining the change in criterion parameters of the semiconductor device under influence of various types of ionizing irradiation. In this case, it is necessary to establish the ratio between damage coefficients for a given type of semiconductor device and its initial characteristics. Therefore, the algorithm of the RadM creation for general case, when there are not the theoretical laws that determine relation between parameters of the semiconductor device and electrophysical with geometrical characteristics of the given semiconductor structures.

These approaches for creating the RadM are fully applicable to develop the RelM of the semiconductor device.

3. Stages of practical implementation of radiation (reliability) model of the semiconductor device creation

The foregoing approach to create the RadM of semiconductor device can be represented as a set of the stages, whose solution of specific problem is characterized. We note that the problem solution of each several stage can already provide practical usefulness.

Subsequently, the RadM will be used as a basis for consideration, because the available experimental data allow to proceed with its development now as it will be show in the following.

On the first stage of RadM development it is necessary to research the change in parameters and characteristics of the semiconductor device under influence of ionizing irradiation.

Such research makes possible to estimate the relations that describe the changing in different parameters and characteristics of the semiconductor device depending on irradiation level. In addition, the damage constants will be determined for various types of ionizing irradiation.

On the next stage of the RadM development it is necessary to relate damage constants of the semiconductor device with its initial parameters. Obtained data in such way have practical importance, because based on them the additional radiation resistance sorting of the manufactured semiconductor devices can be introduced.

At present moment, for instance, the relations have been established that describe the change in majority charge carrier density and charge carrier mobility under ionizing irradiation for device structures based on silicon and gallium arsenide which were manufactured by various technique [8, 22–24]. The initials semiconductor devices can be sorted by radiation resistance following these relations.

The experimental date obtained in this way can be put to the purpose to develop the RadM of Gunn diode, Schottky barrier diode; and Schottky field-effect transistor as is shown in [4, 23, 25].

Next, it is necessary to research the influence of technology and design factors on radiation resistance of the semiconductor device. This can be provided by analysis of manufacturing technology of the semiconductor device for the purpose of detection of the most critical manufacturing process as illustrated in [26] for Gunn diode.

The presented algorithm of development of the RadM of the semiconductor device shows complexity of the model and its development. It is a reminder that during development of the model it is necessary to take into account complex and combined influence of ionizing irradiation and long-term operation factors, which are characteristic of field operating conditions.

This algorithm of development of the RadM of the semiconductor device are fully applicable to develop the RelM of the device with account for specificity of operation factors.

4. Practical implementation of radiation and reliability model of the semiconductor device

Let us take a closer look at practical implementation of RadM and RelM of semiconductor device by the example of the LEDs. In this work we selected industrial IR-LEDs based upon double AlGaAs heteroctructures. Their manufacturing technology, measuring methods of main parameters and using equipment for determination of reliability indexes and for irradiation by fast neutrons and gamma-quanta were described in more detail in [27].

In this paper we confined to the most distinctive types of ionizing irradiation, which were fast neutrons and gamma-quanta. The decision depends on the following motivation. There are two main mechanisms of defect formation under influence of ionizing irradiation on condensed matter. Firstly, it is ionization mechanism of defect formation, which is typical for gamma-quanta, low-energy electrons and protons in fore part of path [21, 28–32]. Secondly, displacement effect of defect formation is typical for fast neutrons, low-energy protons (at the close of particle path) and heavy ions [18, 19, 33, 34].

Therefore, the use of gamma-quanta and fast neutrons makes possible to research the change in parameters and characteristics of the semiconductor device under developing process of two typical mechanisms of defect formation.

We established [35, 36] that the decrease of emissive power of the LEDs under irradiation by fast neutrons and gamma-quanta includes three stages as illustrated in Figure 1.

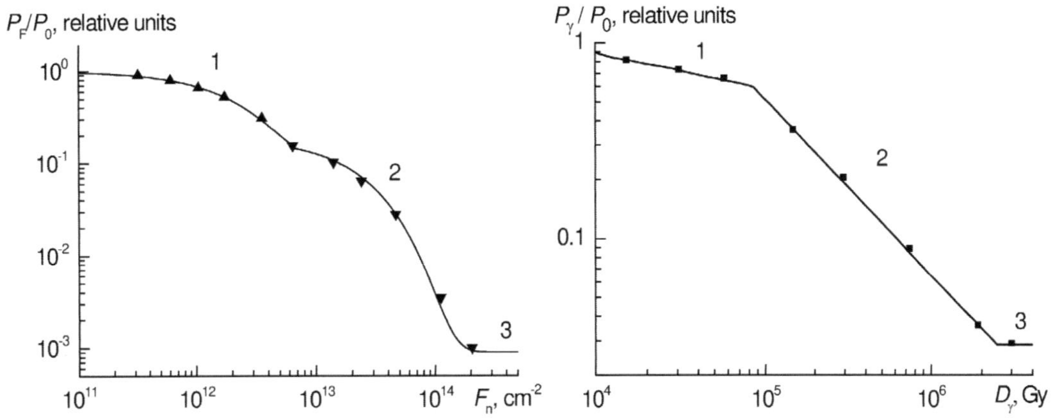

Figure 1. Relative change in the emissive power depending on fast neutron fluence (left) and gamma-quanta dose (right), where 1, 2, 3 are stages of the emissive power decrease.

The above presented research results allow to present the following RadM of the LEDs, which describes changing in emissive power under irradiation by fast neutrons and gamma-quanta.

- On the first stage the decrease of emissive power of the LEDs is attributed to the radiation-stimulated reconstruction of the initial defect structure of the LED crystal as evidenced by saturation of this stage as the level of exposure increases (field 1 in Figure 1).

- On the second stage the decrease of emissive power of the LEDs is attributed to the introduction of new radiation defects (field 2 in Figure 1).

- On the third stage the LED transits into the field of low electron injection in its active region (field 3 in Figure 1). The field is characterized by the fact that the emissive power does not depend on operating current. It should be noted that the boundaries between these stages are rather sharp.

Likewise, the emissive power decrease of the LEDs was researched during their long-term operation. We have been established [37] that there are three distinctive stages of the emissive power decrease of the LEDs during long-term operation referring to Figure 2.

Previously we have been observed the similar behavior of decreasing the emissive power of the identical LEDs under irradiation by fast neutrons and gamma-quanta [35, 36].

Therefore, our results (see Figure 2) can assume the following model for decreasing the emissive power of the LEDs during operation (RelM of the LEDs).

- On the first stage the emissive power decreases due to rearrangement of original defect structure exposed to operation factors in the same way that it was observed under irradiation previously (field 1 in Figure 2).

- On the second stage the emissive power decreases due to introduction of new defects by aging (field 2 in Figure 2).

- On third stage we observe a transition of the LEDs into the mode of low electron injection with further origination of catastrophic failure like the previously observed effect of ionizing irradiation. Moreover, the degradation of ohmic contacts "metal – semiconductor" takes priority of the catastrophic failure (field 3 in Figure 2).

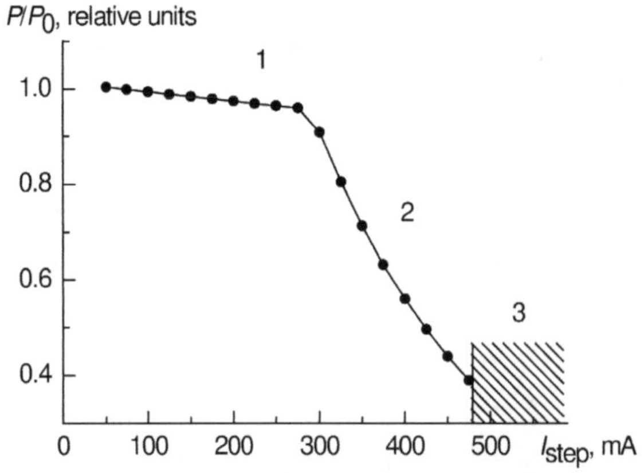

Figure 2. Relative change in the emissive power of the LEDs depending on step number, where 1, 2, 3 are stages of the emissive power decrease.

Crucially, the eigenvalue ratio between established stages are observed for each of influencing factors. For instance, the contribution of the first stage of decreasing emissive power of the LEDs can be estimate at 85% (Figure 1) for fast neutrons. Moreover, it is about 40% from initial value (Figure 1) for gamma-quanta, while the contribution of the first stage is about 5% during operation (Figure 2).

The presented results prove the identity of degradation processes in the LEDs under influence of ionizing irradiation and long-term operation factors. In other words, in this case we can discuss about unify model of the emissive power degradation of the LEDs under ionizing irradiation and during log-term operation. Therefore, the described model is substantially the template for EM-LED.

We note that for creating fully-featured EM-LED it is necessary to provide the research for complex and combined influence of various types of ionizing irradiation and factors of long-term operation.

5. Conclusion

1. The structure of the exploitation model of the semiconductor device is presented, which is based on radiation and reliability models. Furthermore, it was shown that the exploitation model should take into account complex and combine influence of various ionizing irradiation types and operation factors.
2. The algorithm of developing the exploitation model of the semiconductor devices is proposed.
3. The possibility of creating the radiation model of Schottky barrier diode, Schottky field-effect transistor and Gunn diode is shown based on the available experimental data.
4. The basic exploitation model of IR-LEDs based upon double AlGaAs heterostructures is represented.
5. The practical application of the exploitation models will allow to output the electronic products with guaranteed operational properties.

References

[1] Kulakov V M *et al* 1980 *Influence of penetrating radiation on electronic devices* (Sovetskoe radio: Moscow)

[2] Srour J R *and* McGarrity J M 1988 *Proc of the IEEE* **76(11)** 1443-1469 doi: 10.1109/5.90114

[3] Galtseva O V *et al* 2016 *IOP Conf Ser: Mater Sci Eng* **110** 012094 doi: 10.1088/1757-899X/110/1/012094

[4] Plotnikova I V *et al* 2018 *IOP Conf Ser: Mater Sci Eng* **289 (1)** 012029 doi:10.1088/1757-899X/289/1/012029

[5] Plotnikova I *et al* 2018 *MATEC Web of Conferences* **155** 01052 doi: 10.1051/matecconf/201815501052

[6] Galtseva O V *et al* 2015 *IOP Conf Ser: Mater Sci Eng* **81(1)** 012062 doi: 10.1088/1757-899X/81/1/012062

[7] Zeller H R 1995 *Solid-State Electronics* **38(12)** 2041-2046 doi: 10.1016/0038-1101(95)00082-5

[8] Gradoboev A V *and* Surzhikov A P 2005 *The radiation resistance microwave devices based on*

gallium arsenide (Publishing House of Tomsk Polytechnic University: Russia)
[9] Martin D I *et al* 2004 *IEEE Trans Ind Appl* **40(1)** 41-52 doi: 10.1109/TIA.2003.821655
[10] Kapranov B I *and* Varga V V 1999 *Proc of KORUS'1999* **2** 876255 672–676 doi: 10.1109/KORUS.1999.876255
[11] Alkhimov Yu V *et al* 2012 *Russian Journal of Nondestructive Testing* **48(4)** 238–244 doi: 10.1134/S106183091204002X
[12] Surzhikov A P *et al* 2016 *IOP Conf Ser: Mater Sci Eng* **110(1)** 012002 doi: 10.1088/1757-899X/110/1/012002
[13] Gradoboev A V *and* Rubanov P V 2013 *Izvestiya vyzov. Fizika* **56(1-2)** 49-52
[14] Bemš J et al 2014 Radiation Physics and Chemistry 104 398–403 doi: 10.1016/j.radphyschem.2014.02.008
[15] Busatto G *et al* 2013 *IEEE Trans Nucl Sci* **60(5)** pp 3793-3801 doi: 10.1109/TNS.2013.2278038
[16] Liu C *et al* 2015 *IEEE Trans Nucl Sci* **62(6)** pp 3381-3386 doi: 10.1109/TNS.2015.2498201
[17] Schubert E F 2006 *Light Emitting Diodes* (Cambridge University Press: Cambridge)
[18] Burke E A *et al* 1987 *IEEE Trans Nucl Sci* **34** 1220-1226 doi: 10.1109/TNS.1987.4337456
[19] Summers G P *et al* 1988 *IEEE Trans Nucl Sci* **35(6)** 1221-1226 doi: 10.1109/23.25443
[20] Stoneham A M 2001 *Theory of Defect in Solids: Electronic structure of defects in insulators and semiconductors* (Oxford University Press) doi: 10.1093/acprof:oso/9780198507802.001.0001
[21] Vavilov V S *et al* 1981 *Mechanisms of defect formation and migration in semiconductors* (Nauka: Moscow)
[22] Korshunov F P *et al* 1978 *Radiation effects in semiconductor devices* (*Nauka i Tekhnika: Minsk)*
[23] Einspruch N G *and* Wisseman W R (eds) 1985 *GaAs Microelectronics – VLSI Electronics Microstructure Science* **11** (Orlando, Fla.: Academic)
[24] Astvatsaturyan E R *et al* 1988 *Zarubezhnaya elektronnaya tekhnika* **1** 48-83 (in Russian)
[25] Obolenskaya E S *et al* 2015 *Physics and technology of semiconductors* **49(11)** 1507-1515 (in Russian)
[26] Chaffin R J 1973 *Microwave semiconductor devices: fundamentals and radiation effects* (John Wiley and Sons Inc: New York)
[27] Gradoboev A V *et al* 2016 *Microelectronics Reliability* **65** 55-59 doi: 10.1016/j.microrel.2016.07.143
[28] Vavilov V S *et al* 1977 *Radiation effects in semiconductors and semiconductor devices* (Consultants Bureau: New York) doi: 10.1007/978-1-4684-9069-5
[29] Klinger M I *et al* 1981 *Radiation Effects* **56(3-4)** 229-239. doi: 10.1080/00337578108229895
[30] Plotnikova I V et al 2018 IOP Conference Series: Materials Science and Engineering. 289 012029 doi:10.1088/1757-899X/289/1/012029
[31] Lang D V *et al* 1979 *Physical Review Letters* **42(20)** 1353 doi: 10.1103/PhysRevLett.42.1353
[32] Matvienko Y G 2006 *Models and criteria of fracture mechanics* (Fizmatlit, Moscow)
[33] Tapero K I *et al* 2012 *Radiation effects in silicon integrated circuits intended for space application* (BINOM. Laboratoriya znaniy: Moscow)
[34] Gradoboev A V *and* Tesleva E P 2017 *J Phys: Conf Ser* **830** 012133 doi: 10.1088/1742-6596/830/1/012133
[35] Gradoboev A V *and* Orlova K N 2015 *IOP Conf Ser: Mater Sci Eng* **81** 012008 doi: 10.1088/1757-899X/81/1/012008
[36] Gradoboev A V *and* Sednev V V 2014 *Izvestiya vyzov. Fizika* **57(10-3)** 20–23
[37] Gradoboev A V *et al* 2016 *Proc of SIBCON'2016* 1-5 doi: 10.1109/SIBCON.2016.7491662

IOP Conf. Series: Materials Science and Engineering **363** (2018) 012026 doi:10.1088/1757-899X/363/1/012026

Spherical and cylindrical particle resonator as a cloak system

I V Minin, O V Minin, A I Eremeev, I S Tseplyaev
Tomsk State University, Tomsk, Russia
E-mail: prof.minin@gmail.com

Abstract. The concept of dielectric spherical or cylindrical particle in resonant mode as a cloak system is offered. In fundamental modes (modes with the smallest volume correspond to $|m| = 1$, and $s = 1$) the field is concentrated mostly in the equatorial plane and at the surface of the sphere. Thus under resonance modes, such perturbation due to cuboid particle inserted in the spherical or cylindrical particle has almost no effect on the field forming resonance regardless of the value of internal particle material (defect) as long as this material does not cover the region where resonance takes place.

1. Introduction

As it well known, resonant phenomena in cavities, be they acoustic, optical, or otherwise, are frequently dependent on the precise geometric properties, such as size, shape, and composition, of the supporting structure. First explained by Lord Rayleigh [1, 2], whispering gallery modes (WGM) comprise a traveling pressure wave guided around a closed concave surface, such as the whispering gallery in St. Paul's Cathedral [3]. From a geometric optics approximation, such bound modes are guided by means of repeated reflections, which, neglecting absorption, scattering, and material dispersion, continue ad infinitum. This fact has motivated us to consider a spherical particle resonator as a cloak system.

It could be noted that the single subwavelength defect-induced backscattering in WGM resonator was investigated in [4, 5, 6] and was experimentally confirmed in [7]. In [8] a connection between the interaction of WGMs with a single subwavelength defect and the formation of the spectral doublets was directly confirmed.

Today, the two most developed cloaking techniques are: cloaking based on the wave flow method and cloaking based on scattering cancellation [9]. In the first method the object is placed in a shell which causes electromagnetic fields to bend around the object and to recover their wavefront and intensity distribution afterwards. The operation of such a shell does not depend on the properties of the hidden object, because in this case electromagnetic waves do not interact with the object.

2. Modeling

Let's briefly consider the interaction of electromagnetic field with cylindrical dielectric particle in resonant mode. An infinite long transparent dielectric square section cuboid embedded in a infinite long transparent dielectric cylinder is illuminated by a monochromatic linear polarized plane wave with an incident wave vector parallel to the Z axis. Its polarization direction is parallel to the axis of the cylinder.

We simulated this combined structure using Finite Element Method (FEM) software Comsol Multiphysics. Perfectly matched layers (PML) bound the simulation space at the all sides. At the cylindric's shadow side, additional space with the width 2λ is added to visualize and study the resulting modes. We used a free tetrahedral mesh with maximum element size $\lambda/5$ for the free-space regions, and $\lambda/5/1.8$ for the particle. Additionally, we used a free triangular mesh with maximum element size $\lambda/20$

for the z-plane near the shadow surface of cylinder (sphere) and an even finer mesh is required for obtaining the field's spatial structures.

The simulated field magnitudes are shown in Figure 1 for cylinder diameter of 6.6234 λ and refractive index of n=1.75. In the Figure1a the field distribution inside and outside the homogeneous cylinder are shown. In the Figure 2b the field distribution inside and outside the homogeneous cylinder the case when a cuboid dielectric particle with dimensions of 2.41λ × 2.41λ and with a refractive index of $n_2 = 1.9$ placed at the center of the cylinder is plot. As it can be seen the effect of the cuboid on the scattered field from the cylinder is clearly observed; and the resonance behavior remains unchanged. In fundamental modes (modes with the smallest volume correspond to $|m| = 1$, and $s = 1$) the field is concentrated mostly in the equatorial plane and at the surface of the sphere. Thus under resonance modes, such perturbation due to cuboid has almost no effect on the field forming resonance regardless of the value of internal particle material (defect) as long as this material does not cover the region where resonance takes place.

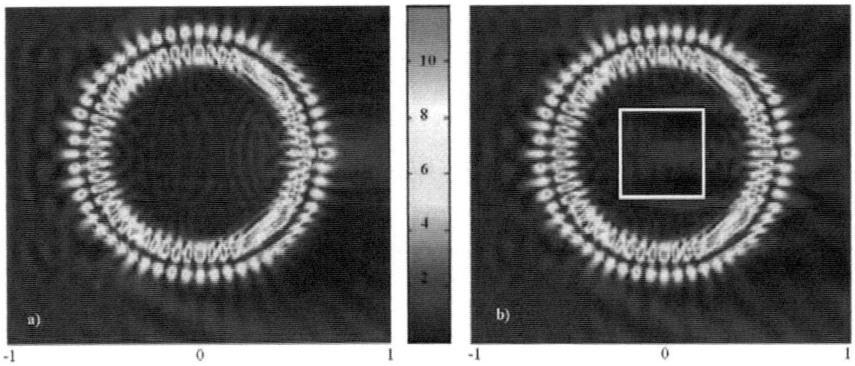

Figure 1. The simulated field magnitudes for dielectric cylinder in resonance mode.

The correspondent field magnitudes to the cases shown in Figure 1 along the cylinder diameter are shown in the Figure 2.

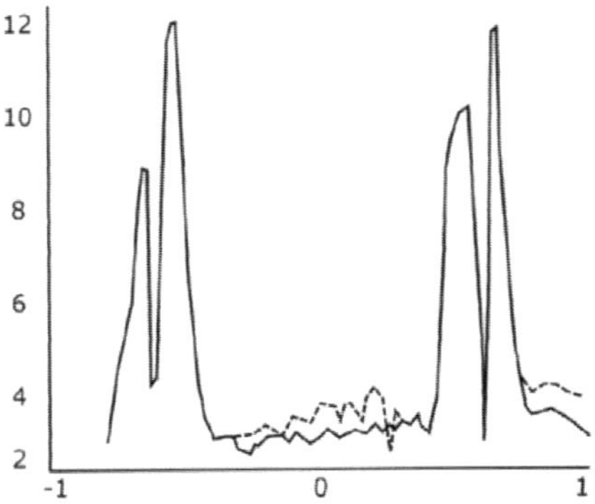

Figure 2. Magnitude of the total field along the diameter of the cylindrical particle: solid line corresponds to the homogeneous cylinder and the dashed line corresponds to the case when a cuboid dielectric particle placed at the center of the cylinder.

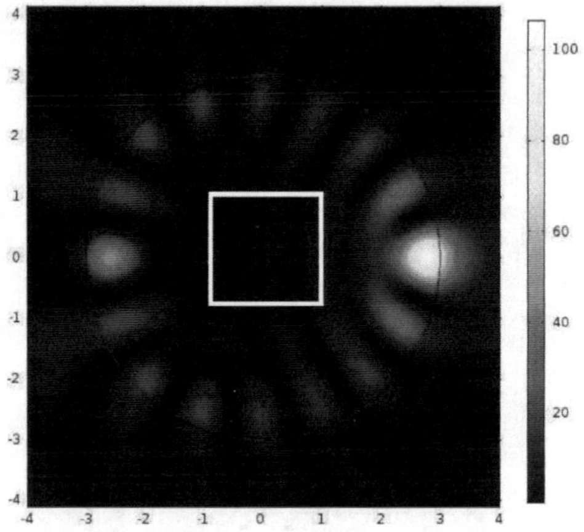

Figure. 3. The simulated field magnitudes for dielectric sphere in resonance mode.

The same effect was observed for spherical particle shown in the Figure 3. In other words, if the above conditions are met, the cubic particle will not be visible when analyzing the external field of the spherical or cylindrical dielectric particle resonator.

Additional simulations shown that in non-resonant modes if we placed cuboid particle near the shadow surface inside of cylindrical (spherical) particle a photonic jet [10] focused outside the cuboid with the beam waist (FWHM) about two times smaller than that generated by the homogeneous dielectric cylinder. This effect is analogous to the effect of the photonic jet formation on the basis of a cubic particle irradiated by a focused beam [11-12].

Example of such beam compressed system concept, based on dielectric self-similar cuboids, was described in [11]. Initial wave beam is focused and transported by dielectric cubic particles with the size, magnified by M times (this can be a conventional lens system or a single lens). A mesosize dielectric cube with minimal optimal sizes, is placed in the focus of this system. Then, the initial wave beam is focused into a subwavelength region with a transverse dimension less than half a wavelength.

3. Conclusion

So, the capability of the flat reflection array to focus a monochromatic radiation at a certain point were studied. The element of the array based on waveguide with a controlled reflection coefficient was developed. The controlled phase shift is 180°. The array model based on waveguide elements was simulated in CST Microwave Studio. These simulations proved the monochromatic radiation focusing with the array made up of waveguide elements. Experimental studies of a reflective array consisting of 100 waveguide elements confirmed the focusing properties of such a array. A hardware and software complex to control the array is being currently developed. The proposed structure of the reflection array provides the high-rate focusing at a certain point.

Acknowledgments

The research was supported by the Russian Ministry of Education and Science as a part of the state order No 3.2068.2017/4.6.

References

[1] Rayleigh L 1910 *Philos. Mag.* **20** 1001-1004
[2] Rayleigh L 1910 *Philos. Mag.* **27** 100-109
[3] Raman C V and Sutherland G A 1921 *Nature* **108** 42
[4] Little B E, Juha-Pekka Laine J P and Chu S T 1997 *Opt. Lett.* **22** 4-6
[5] Gorodetsky M L, Pryamikov A D and Ilchenko V S 2000 *J. Opt. Soc. Am.* **17** 1051-1057

[6] Borselli M, Johnson T, and Painter O 2005 *Opt. Express 13* 1515-1530
[7] Kippenberg T J, Spillane S M and Vahala K J 2002 *Opt. Lett.* **27** 1669-1671
[8] Mazzei A 2007 *Phys. Rev. Lett.* **99**, 173603
[9] Fleury R, Monticone F and Alù A 2015 *Phys. Rev.* **4** 037001
[10] Luk'yanchuk B S, Paniagua-Domínguez R, Minin I V, Minin O V and Wang Z 2017 *Optical Materials Express* **7** 1820-1847
[11] .Minin I V, Minin O V, Nefedov I, Pacheco-Peña V and Beruete M 2016 Beam compressed system concept based on dielectric cluster of self-similar three-dimensional dielectric cuboids. *Proc. of the Int. Joint Conf. GSMM 2016* (Espoo Finland) 64-66
[12] Pham H H N, Hisatake S Minin O V, Nagatsuma T and Minin I V 2017 *APL Photonics* **2** 056106

System of technical vision for autonomous unmanned aerial vehicles

A S Bondarchuk
Student, National Research Tomsk State University, Tomsk, Russia
E-mail: bondarchuk.a.c@gmail.com

Abstract. This paper is devoted to the implementation of image recognition algorithm using the LabVIEW software. The created virtual instrument is designed to detect the objects on the frames from the camera mounted on the UAV. The trained classifier is invariant to changes in rotation, as well as to small changes in the camera's viewing angle. Finding objects in the image using particle analysis, allows you to classify regions of different sizes. This method allows the system of technical vision to more accurately determine the location of the objects of interest and their movement relative to the camera.

1. Introduction

One of the important ways to use unmanned aerial vehicles (UAVs) is to collect information. UAVs are used in various areas of human activity: security monitoring, geodesy, exploration, etc. The greatest amount of information is contained in the video signal [1, 2]. To work with this information, UAVs are equipped with vision systems, and image storage and transmission system for later use for other purposes. With the help of a vision system, unmanned aerial vehicles will be able to recognize and respond to certain objects. This system will allow the UAV to follow a certain object and thereby monitor its location. This paper was devoted to creating a virtual instrument in LabVIEW for detecting objects on the frames from the camera mounted on the UAV.

2. Classifier training

The image of an object can be directly represented as a vector in a multidimensional space. Then train some classifier using a set of sample images of objects. The classifier in this case is some tool that takes an input image, represented as a vector in a multidimensional space, and outputs some information that classifies the input image relative to some attribute [3, 4].

To train the classifier, the virtual instrument "HOG-LBP Texture Classification" was used, which is located in the LabView Examples. The IMAQ Extract HOG Feature Vector and IMAQ Extract LBP Feature Vector blocks are combined to form a feature vector. The best results were obtained using the following parameters: Grid Size: 4×4 (along the x and y axes respectively); Histogram Bins: 25.

The classifier was trained in two classes, the class "Car", in which samples of vehicle images were added, and the class "negative". In the "negative" class, samples that did not contain objects of interest were added. When the first sample was added, the SVM model of one class was used, the C-SVC model was used for the subsequent training.

3. Pattern recognition on frames from drones

This virtual instrument is designed to detect vehicles on an image or video stream from a camera mounted on a UAV. To begin with, Vision Acquisition VI was installed, After closing the configuration window, a While loop is created in the VI, which contains the Vision Acquisition and Image Out blocks. Next, the construction of the entire block diagram will occur within this cycle.

The frames received with the Vision Acquisition block will be delivered to the IMAQ ExtractSingleColorPlane and to the Vision Assistant. The frames arriving at the input of this module are converted first into images in gradations of gray (with a sensitivity to the red color). Further on the received images the filter using edge detection-prewitt is superimposed.

After using edge detection-prewitt, a threshold binarization of the image is performed [5, 6]. When setting the binarization parameters, the "Auto Threshold: Clustering" method was selected and the Lower Limit value was set to 10 and the Upper Limit was set to 245. In the Look For section, the Bright Objects option was chosen to assign the value 1 to the pixels at the object boundary. The last image conversion in Vision Assistant is the use of the Equalize function, which assigns the values 0 and 255 to the pixels 0 and 1. All these transformations are necessary for further work with the images, namely finding the coordinates of regions containing a continuous region of non-zero pixels. These areas are usually called particles.

A subVI was created to detect pixel groupings in the image and sequentially obtain the coordinates of these groupings. Particle analysis consists of a series of operations for processing and analyzing functions that provide information about particles in an image. Using particle analysis, you can detect and analyze any shapes in the image. A typical particle analysis process scans the entire image, detects all particles, and builds a detailed report on each particle. You can use several parameters such as perimeter, angle, area, and center of mass to identify and classify these particles. This analysis is necessary for the search for particles whose spatial characteristics satisfy certain criteria. To reduce the calculation time, particle filtration is used, excluding particles that are not of interest based on their spatial characteristics. Only the particles of the necessary size remain for further analysis. Thus, the output of this subVI will be the location of the objects in the image. A block diagram of the virtual instrument Classification analysis is shown in Figure 1.

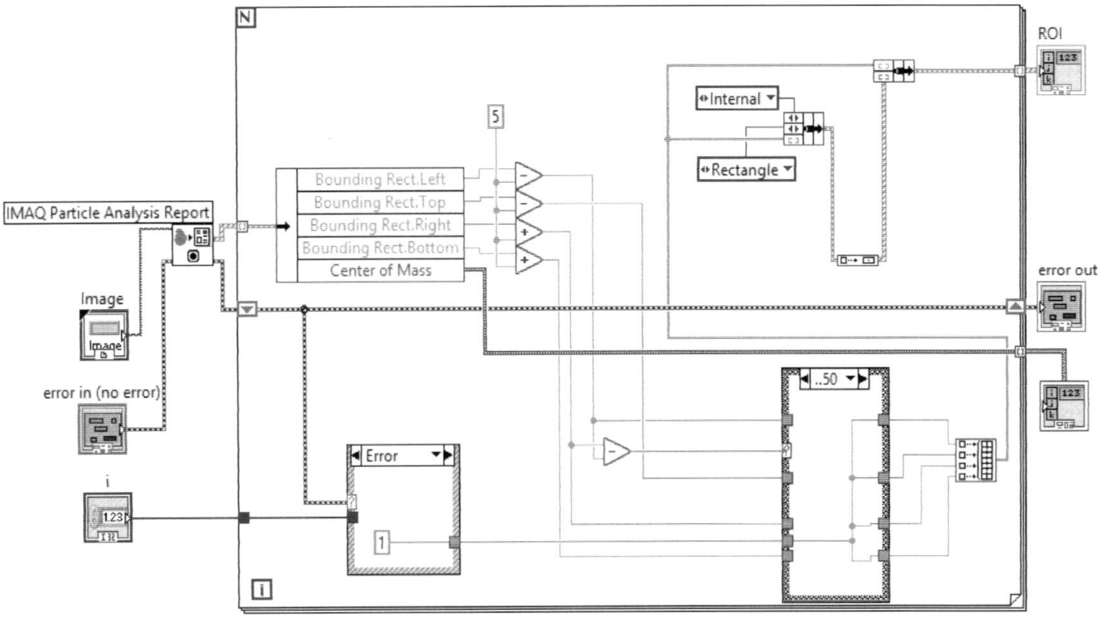

Figure 1. Block diagram of the virtual instrument Classification analysis.

Images are delivered to the IMAQ Particle Analysis Report block, which returns the number of particles found in the binary image and an array of reports containing the most commonly used particle measurements.

IOP Conf. Series: Materials Science and Engineering 363(2018) 012027 doi:10.1088/1757-899X/363/1/01202

Subsequently, in the sub-VI Classification analysis, only the Particle Reports cluster (Pixels) will be used. To process data for all image particles, the For Loop is used.

The Particle Reports cluster (Pixels) uses the Unbundle By Name function, which returns cluster elements such as Bounding Rect and Center of Mass. To the coordinates of the sides of the rectangle, 5 pixels are added in order to have additional space around the particle, and it does not touch the boundary of the ROI. These coordinates are then fed to Case Structure. The Case Selector input receives the rectangle width values for each particle. If the width of the rectangle surrounding the particles is less than 50 pixels, the values of the coordinates of its sides become equal (this value varies depending on the iteration cycle). If the width of the rectangle is more than 50 pixels, the coordinate values are not changed. This is done in order to reduce the time spent processing the image, since objects of such small size are not of interest. Equal values of the coordinates of the sides of the rectangle can cause an error in the work of the VI. This is because the IMAQ Extract HOG Feature Vector and IMAQ Extract LBP Feature Vector blocks can not form a feature vector from this region. To prevent such an error, use Case Structure, whose input receives an error. In the absence of errors, the structure sums the unit and the current value of the iteration cycle, which is fed to the input of sub-VI Classification analysis. The resulting value goes to the previous Case Structure, where it replaces the coordinate values of rectangles that are less than 50 pixels wide. When an error occurs, this value becomes equal to one.

From coordinates passed through Case Structure, a data array is built using the Build Array block. This array is fed to the Bundle function, which creates a cluster of three elements (the array is fed to the input, for the third element of the cluster). To the input, for the first element, is fed by Enum Constant. By default, the numeric representation of this constant is a 16-bit unsigned integer. To configure the constant, after clicking on it with the right mouse button, select the Properties section. In the appeared window, in the Appearance tab, in the Label section, the word ID is written. In the next Data Type tab, the numeric representation of this constant is changed to a 32-bit unsigned integer. In the Edit Items tab, External marks are added with a value of 0 and Internal with a value of 1. After clicking OK, the Internal label is selected in the constant. On the input, for the second element of the cluster created with the Bundle function, another Enum Constant is supplied. When you configure it in the Appearance tab, in the Label section, the word Type is written. The Data Type is set to 32-bit, and the Edit Items are added with the labels specified in Table 1. After clicking OK, the Rectangle label is selected in the constant.

Table 1. Tags added to Edit Items.

Tag	Value
No Selection	0
Point	1
Line	2
Rectangle	3
Oval	4
Polygon	5
Free	6
Unused1	7
Unused2	8
Unused3	9
Broken Line	10
Free Hand Line	11

The resulting cluster of three elements is converted into a data array using the Build Array block. This array is fed to the Bundle function, as the second element, of the new cluster being created. To the input for the first element is fed an array of coordinates, which was used in the creation of the previous cluster. The new cluster created in this way represents the regions of interest (ROI). In these regions, there are objects that will be classified to search for vehicles in the image.

So, sub VI Classification analysis was created. Its input receives the image and the value of the iteration cycle (i). At the output, the regions of interest and the centers of mass of the image particles are returned.

For further classification, the For Loop cycle was created. Within this cycle, an ROI cluster and a pre-converted image in grayscale (block IMAQ ExtractSingleColorPlane, in the parameter Color Plane is selected Red) are delivered to the inputs of the IMAQ Extract HOG Feature Vector and IMAQ Extract LBP Feature Vector blocks. In the parameters extraction features from the image, the Histogram Bins value is set to 25 and the Grid Size is 4 × 4. Next, the generated feature vector is fed to the IMAQ Classify Custom block, which classifies this vector using the classifier (which is also fed into the input).

Also, another subVI was created, which eliminates unnecessary information about objects that are not of interest for the developed system of technical vision. A block diagram of the virtual Instrument Condition is shown in Figure 2.

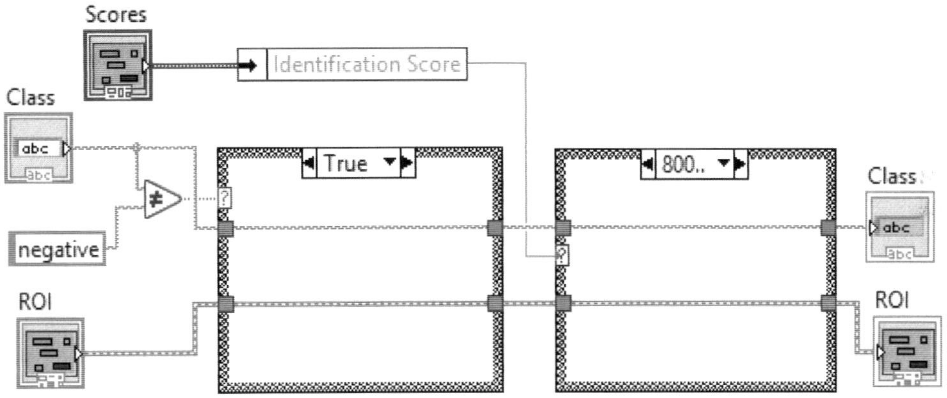

Figure 2. Block diagram of the virtual instrument Condition.

The classifier of the technical vision system was trained in two classes, the class "Car", which includes attributes belonging to vehicles, and the class "negative", representing samples of other objects. In order to distinguish objects of only one class (namely "Car"), Case Structure was built. Its input receives a variable of type Boolean, which takes the value "False" if the sample is classified as "negative". Thus, if the input is set to "True", the data for the classification object passes through Case Structure. If the input is set to "False", the data is assigned zero values by means of creating constants. Through this structure, we direct the Class variable obtained at the output of the IMAQ Classify Custom block and the ROI that were obtained at the output of the sub-VI Classification analysis.

Further, these variables must pass through yet another Case Structure, which sets a valid value for the score of the matching of input samples to the assigned class. This value ranges from 0 to 1000, where 1000 represents 100% certainty of belonging to the class. The input of this structure receives the Identification Score variable, obtained by using the Unbundle By Name function on the Scores cluster (obtained from IMAQ Classify Custom). If the value of this variable is greater than 800, the data about the classified objects pass through the Case Structure without changes. For smaller values, the data is assigned zero values.

As a result, Class, Scores and ROI variables are supplied to the input of the Sub VI Condition, and the output returns the values of the variables (Class and ROI) that satisfy the specified requirements. The IMAQ Overlay ROI and IMAQ Overlay Text blocks are used to display the areas of interest and class names. The input of these blocks receives the variables ROI, Class and frames from the drone's

IOP Conf. Series: Materials Science and Engineering 363 (2018) 012027 doi:10.1088/1757-899X/363/1/01202

camera. These frames can be both original images from the camera, and pre-filtered, it depends on what image needs to be obtained at the output of the VI.

It should be noted that for the IMAQ Overlay Text block, coordinates are also needed, according to which the text will be displayed. This can be the coordinates of the center of mass obtained in the sub-VI Classification analysis. Figure 3 is a block diagram of an implemented virtual instrument intended for recognizing vehicles on frames received from cameras installed on a UAV.

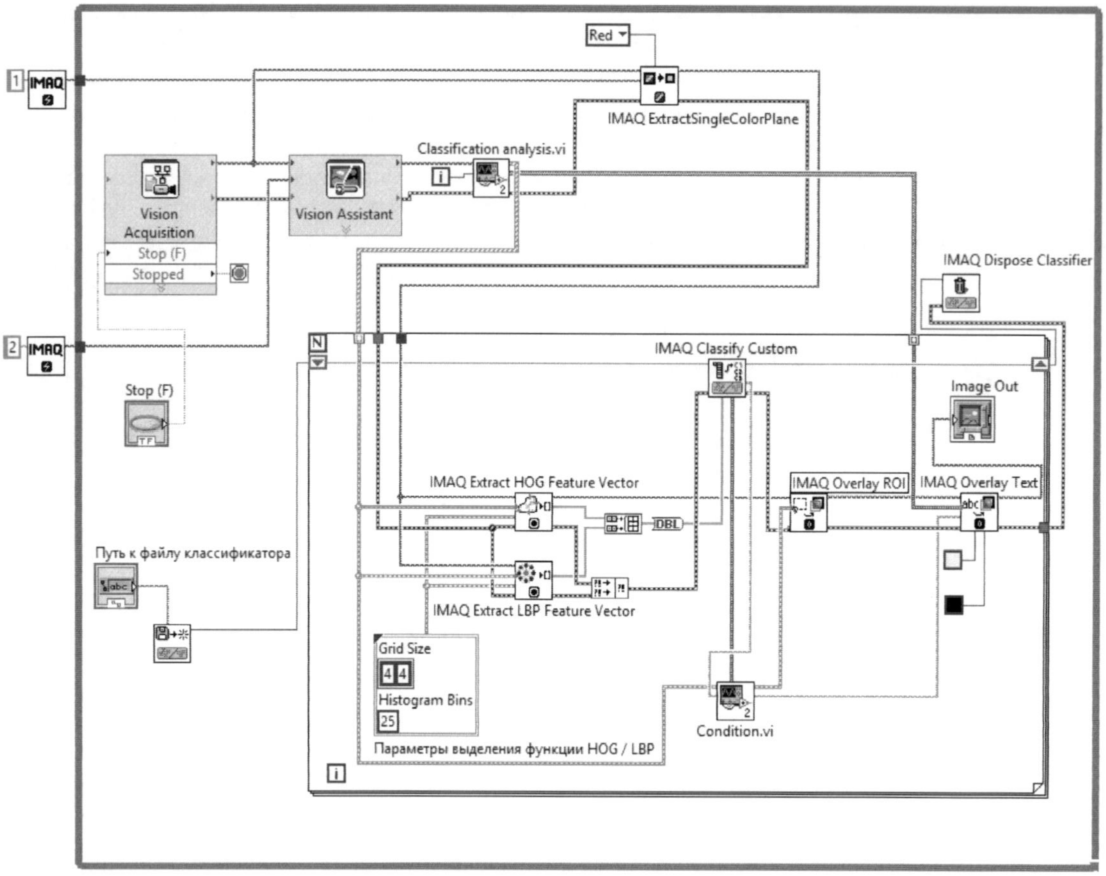

Figure 3. Block diagram of VI intended for pattern recognition on frames from UAV.

4. Results

Finding objects in the image using particle analysis, allows you to classify regions of different sizes. This method allows the system of technical vision to more accurately determine the location of the objects of interest and their movement relative to the camera. The classifier, trained in vehicle recognition, is invariant to changes in rotation, as well as to small changes in the camera's viewing angle.

The disadvantage of the method using particle analysis is that when changing lighting or background environments, there is a chance that object in the image will not be found. This is due to the change in the contrast of the image, so that the outlines of objects may not be highlighted during filtering. For example, a car, painted in black, drove into a section of the road that is in the shade of a large tree. In such a situation, the difference in the pixel values of the object and its background is small, and consequently, obtaining of the boundaries of the object is problematic. When the contrast of the image is increased with the help of filters, the number of particles in the image increases, which hinders the determination of the size of objects and their location. Also, increasing the contrast adds extra noise to the image.

Consequently, this virtual instrument most effectively distinguishes objects of interest on areas whose color palette is monotonous, and differs from the color palette of objects.

The work was carried out in Tomsk State University with the financial support of the Ministry of Education and Science of Russia, the unique identifier of the project RFMEFI57817X0241.

References

[1] Shashev D V, Shidlovskiy S V, Syriamkin V I, Yurchenko A V 2015 *IOP Conference Series: Materials Science and Engineering* **81** 012101
[2] Panin S V, Syryamkin V I, Glukhikh A I 2003 *Avtometriya* (4) 79–92
[3] Yane B 2007 *Digital image processing* (Moscow, Technosphere)
[4] Okhotnikov V V *et al* 2016 *IOP Conference Series: Materials Science and Engineering* **116** 012001
[5] Shidlovskiy S 2016 *MATEC Web of Conferences* **79** 01014
[6] Shashev D, Shidlovsky S 2017 *Journal of Physics: Conference Series* **881(1)** 012029

IOP Conf. Series: Materials Science and Engineering **363** (2018) 012028 doi:10.1088/1757-899X/363/1/012028

Realization of the FPGA-based reconfigurable computing environment by the example of morphological processing of a grayscale image

V Shatravin, D V Shashev
National Research Tomsk State University, Tomsk, Russia
E-mail: dshashev@mail.ru

Abstract. Currently, robots are increasingly being used in every industry. One of the most high-tech areas is creation of completely autonomous robotic devices including vehicles. The results of various global research prove the efficiency of vision systems in autonomous robotic devices. However, the use of these systems is limited because of the computational and energy resources available in the robot device. The paper describes the results of applying the original approach for image processing on reconfigurable computing environments by the example of morphological operations over grayscale images. This approach is prospective for realizing complex image processing algorithms and real-time image analysis in autonomous robotic devices.

1. Introduction

The ubiquitous introduction of information systems in the technological process control requires higher accuracy and operation speed of individual subsystems. As an example, we can take a wide class of data processing tasks generated by a vision system (VS), which provides control and management of technological operations. One of the main data operation is filtering and selection of areas of interest on each frame of the video stream.

The classical processing method implies transfer of scenes digitized by photosensitive matrices to a computer system that performs certain operations in the context of the problem being solved. In this connection, two approaches are used to increase the operation speed:

• increase of the computer system performance;
• improvement and optimization of the algorithms used.

The weak point of this method is serial transfer of information about each pixel of the image from the sensitive matrix to the computer system, which is a constraint that restrains achievement of the greatest performance.

The alternative approach proposed in [1] consists in direct combining sensory (photosensitive matrix) and computational (specialized homogeneous structure) systems in the form of separate layers on a single semiconductor crystal. This will allow transferring data from each matrix sensor to a separate structure calculator, which should lead to a significant increase in the system performance [2]. The main advantages of the approach include the following:

• information from each matrix sensor is processed by a separate calculator, which implies a high degree of parallelism in image processing;

• the speed of such a system does not depend on the number of sensors; it is possible to effectively scale the system without significant loss of speed, which is a weak point in systems with serial data processing.

Within our project, the homogeneous FPGA-based computing structure for morphological processing of grayscale images was realized.

2. Homogeneous computing structure for image processing

The homogeneous structure under consideration is a reconfigurable computing environment (RCE, homogeneous environment) representing a geometrically regular lattice with at least two symmetry axes (Figure 1) and single-type functional cells (calculators) at the nodes. Each functional cell is connected to neighboring cells geometrically in the same way and can be configured to perform a specific signal processing function at a given time. A signal from any cell *A* can be transmitted to any cell *B* of the environment (at least by means of other cells). This homogeneous medium is isotropic, which allows realizing any given function in any part or region of the environment [2].

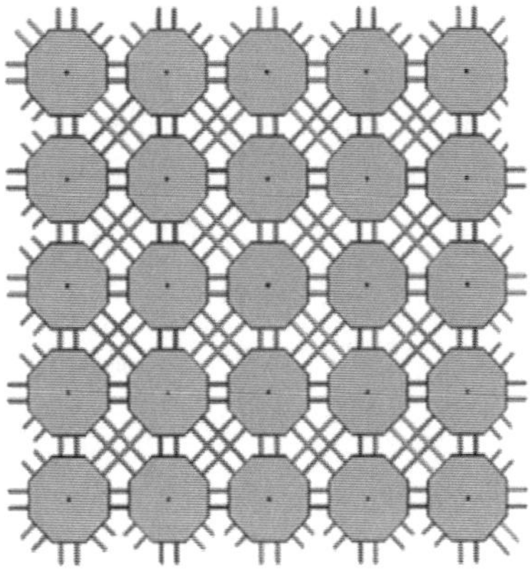

Figure 1. Reconfigurable computing environment.

Each calculator has a specialized structure (Figure 2) and serves to process information about one pixel of the image in the streaming mode. It allows avoiding overhead costs for intermediate transformations and serial data transmission, which are typical of the classical approach. Each calculator receives signals from neighboring calculators and from its "own" sensor of the photosensitive matrix and transmits the output signal to the neighboring calculators and the corresponding pixel of the resulting image. The operating mode of a specific calculator is set by the configurable signal, which allows it dynamic adjusting and influencing the overall processing process.

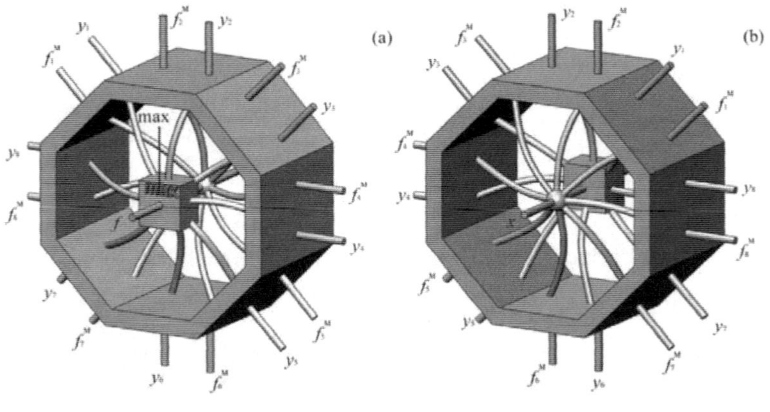

Figure 2. Elementary calculator.

A mathematical apparatus has been developed for implementing the basic morphological operations on binary and grayscale images. Mathematical morphology allows extracting certain image components to represent and describe the shapes of objects. In morphological processing, a binary image represents a set of pixels of the foreground and background; therefore, operations performed on sets can also be performed on images.

Dilation (expansion), erosion (compression), opening and closing are the basic operations in morphological image processing. They are performed on two images: the image being processed and a special image depending on the type of operation and the task being solved. A special image is called a structural element or a mask. A mask is a matrix with a central element (i.e. with an odd number of rows and columns), which defines the form of image processing. A key element is set for each mask; the central element often becomes a key element. In this approach, only the simplest 3x3 structural element is used, all cells of which are filled with “1”.

Morphological operations realized by the calculator include dilation and erosion. More complex operations (for example, opening and closing) can be performed through the implementation of a multilayer computational structure (Figure 3), or through the implementation of push-pull processing, in which the result of one operation is fed to the input of the same calculator, but the calculator itself is tuned to performing another operation.

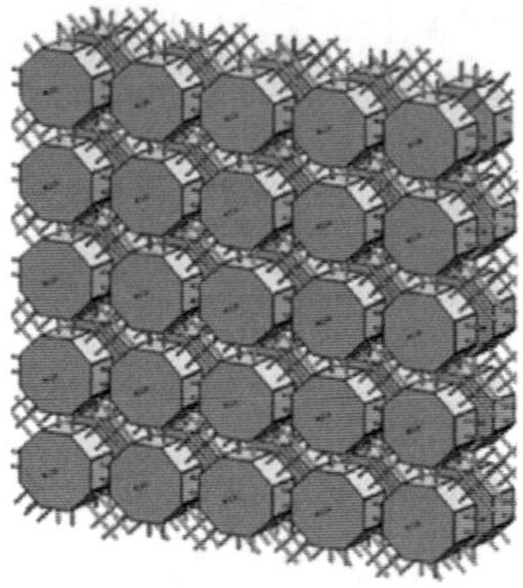

Figure 3. Multilayer computing structure.

This device is sufficient for the image pre-processing (reducing noise, smoothing contours, removing “gaps”, and detecting zones of interest) [3]. Because the processing of grayscale and full-color images will be performed according to the similar algorithms, you will only need to scale the structure for processing a larger data stream.

Because the morphological operations are essentially spatial filters, when processing the boundary pixels of the image, the problem of missing pixels occurs. This problem is solved by introducing a tuning signal. This signal allows encoding one of nine possible states of the neighborhood of the pixel being processed (for a 3x3 mask) including four angular, four edge and one not boundary state. In addition, this signal determines the operation performed by the computing element [1, 4, 5].

3. Selection of a hardware platform for the computational structure implementation

A single element of such a computational structure can be effectively implemented using field programmable gate array (FPGA). The choice of FPGA as a hardware platform is connected with the following factors.

• Similarity. The FPGA structure is similar to the homogeneous environment being implemented; the FPGA structure is a homogeneous matrix of programmable logic blocks connected by programmable internal connections. By recording the operations of morphological processing considered in this task in the form of logical operations, it is possible to implement the logic of calculators and connections between them based on FPGA.

• Parallelism. Logical blocks and their unions do not depend directly on each other; it allows implementing a large number of parallel computational elements on one circuit.

• Reconfigurability. Based on FPGA, it is possible to implement reconfigurable computing structures that can effectively solve complex technical problems.

• High operating speed. At the execution stage, FPGA is a certain digital device, which generally does not read commands and data from the memory, therefore the operating speed of the resulting device depends only on the actuation of logic keys and parasitic reactive processes, which provides high performance and wide optimization possibilities.

• Flexibility. FPGAs have a large number of possible rewriting cycles, which allows quick required correction of the existing configuration and continuation of operation.

• Accessibility and convenience.

As a design environment for the FPGA, we selected Quartus II 13.0 Web Edition, the simulation environment was Altera ModelSim 10.1, the language was Verilog HDL, and a programmable logic device (PLD) was Cyclone IV (EP4CGX110DF31C7).

4. Results

The first step in solving the problem was implementation of a separate pixel-processing calculator for a grayscale image in accordance with the basic operations of morphological processing. This calculator was implemented based on the research [4]. The key difference is processing of not single signals, but 8-bit buses (8-bit numbers), and complication of the logic of the morphological operations (the dilation operation corresponds to the maximum of the current neighborhood and the erosion operation corresponds to the minimum).

Using the module generating function of the Verilog language, a 100×100 matrix and all necessary connections between the elements were obtained. The initial image is loaded as a 100x100x8 bus (length, width, bit width) of binary signals using the $readmemh command. In this regard, the initial image must be a two-dimensional array, each element of which is an 8-bit luminance value of the corresponding pixel. In this format, images with the extension .pgm (portable grayscale map) are represented.

To simulate the resulting computational structure, a grayscale image of 100×100 pixels was selected as shown in Figure 4. For later processing, it was converted to pgm format followed by deleting the file metadata. The resulting image was processed using the obtained computing environment in the simulation mode. Figures 5, 6 show the results of dilation and erosion. Analysis of the timing diagram of the simulation in Quartus showed that the formation of the resulting image requires no more than 21 nanoseconds from the moment of the signal change at the input of the computing environment.

Figure 4. Initial image.

Figure 5. The result of dilation.

Figure 6. The result of erosion.

5. Conclusion

Solution of this problem resulted in implementation of a computer system based on RCE processing of arbitrary sized grayscale images. This FPGA-based computer system is well scalable and allows efficient parallelization of the spatial image processing process between simple calculators performing basic operations of mathematical morphology. Simulation of processing of a grayscale 100×100 pixels 8-bit image was performed in Quartus II 13 environment using the ModelSim 10 simulator. Analysis of the obtained results showed that the implemented structure solved the task correctly and required no more than 21 nanoseconds for forming the resulting image when simulating for FPGA Altera Cyclone IV (EP4CGX110DF31C7).

Acknowledgments

The research was carried out in Tomsk State University with the financial support of the Ministry of Education and Science of the Russian Federation, a unique project identifier: RFMEFI57817X0241.
The authors are grateful to Tatiana B. Rumyantseva from Tomsk State University for her assistance in preparing the paper.

References

[1] Shashev D and Shidlovskiy S 2015 *Optoelectronics, Instrumentation and Data Processing* **51(3)** 19–26 doi: 10.3103/S8756699015030036
[2] Shidlovskiy S 2016 *MATEC Web of Conferences* 79 1–5 doi: 10.1051/matecconf/20167901014
[3] Gonzalez R, Woods R 2009 *Digital Image Processing* (London, Pearson Education)
[4] Shashev D, Shidlovskiy S., Syriamkin V, Yurchenko A 2015 *IOP Conference Series: Materials Science and Engineering* **81(1)** 012101 doi:10.1088/1757-899X/81/1/012101
[5] Yurchenko A, Syriamkin V, Okhorzina A, Kurkan N 2015 *IOP Conference Series: Materials Science and Engineering* **81(1)** 012097 doi:10.1088/1757-899X/81/1/012097

IOP Conf. Series: Materials Science and Engineering **363** (2018) 012029 doi:10.1088/1757-899X/363/1/012029

Understanding of Android-Based Robotic and Game Structure

A Phongtraychack[1,2], V Syryamkin[1]
[1]National Research Tomsk State University, Tomsk, Russia
[2]National University of Lao PDR, Vientiane, Lao PDR
E-mail: rtb98@mail.ru

Abstract. The development of an android with impressive lifelike appearance and behavior has been a long-standing goal in robotics and a new and exciting approach of smartphone-based robotics for research and education. Recent years have been progressive for many technologies, which allowed creating such androids. There are different examples including the autonomous Erica android system capable of conversational interaction and speech synthesis technologies. The behavior of Android-based robot could be running on the phone as the robot performed a task outdoors. In this paper, we present an overview and understanding of the platform of Android-based robotic and game structure for research and education.

1 Introduction

In recent years, androids have become increasingly the popular media and visible device. Android is a Linux based operating system; it is designed primarily for touch screen mobile devices such as smart phones and tablet computers. The operating system (OS) has been improved a lot in last 15 years starting from black and white phones to recent smart phones or mini computers. Nowadays, one of the most widely used mobile OS is android. The android is software that was founded in Polo Alto of California in 2003 [1].

1.1 System Integration

Smart phones and cloud computing were introduced to automation industry as components of connected enterprise integration system. The established protocols and interfacing done in that area in addition to the rapid growth of the hardware specification of those devices were behind the motivation to think of smart phones as programming and simulation devices.

1.2 Related work

Human-robot collaboration has been a significant research topic since the beginning of robotics. The constant introduction of robots in industrial environments, the creation of new compliant robots and the sensors available nowadays in the market (cheaper and more accurate) make human-robot interaction an even more active and exciting research subject [2].

1.3 Humanoid robot

A humanoid robot is a robot with its body shape built to resemble the human body. The design may be for functional purposes such as interacting with human tools and environments, for experimental purposes such as the study of means of locomotion or for other purposes. In general, humanoid robots have a torso, a head, two arms, and two legs; though some forms of humanoid robots may model only part of the body, for example, from the waist up. Some humanoid robots also have heads designed to replicate human facial features such as eyes and mouths. Androids are humanoid robots built to resemble humans aesthetically [3].

2 Platform Architecture

2.1 ERICA platform

There are many variations of Japanese robotics. Erica is an android developed as a research platform for the autonomous conversational robot (Figure 1). Developing total technology to enable Erica to have natural interaction with persons by integrating various technologies such as voice recognition, human tracking and natural motion generation. It has nineteen degrees of freedom for face, neck, shoulder, and waist; it can express various facial expressions and some gestural motions. Its appearance is designed for beautiful and neutral female face, by which people can familiarly interact with it. It speaks in synthesized voice [4].

Figure 1. Autonomous conversational robot Erica.

Currently, the project is in an early phase, so Erica's high-level conversational capabilities are still quite limited. However, the robot platform has been completed, and Erica was recently unveiled in a public demonstration showcasing the basic functionalities of the platform. During the demonstration, it autonomously answered questions from reporters and conducted conversations while demonstrating a rich set of nonverbal behaviour.

2.2 Android based robotic platform

The android based robotics platform was originally developed by Nicolas Oros and Jeffrey Krichmar. Android-based robotics provides an inexpensive platform for robotic control coupling the powerful capabilities of Android smartphones with off-the-shelf robotic components. Examples of its use include academic research in cognition and robotics, robotics competitions, classroom instruction, and personal project (Figure 2) [5].

Figure 2. Android Based Robotic platform.

2.3 Hardware and software

- **Hardware**

An important advantage of using a smartphone for an on-board computer is that the size of a robot can be kept relatively small while preserving great features. Its cost can also be minimal since the phone itself can handle computation, sensing and battery power.

Many different phones are now available on the market. Before purchasing an Android phone to be used as an on-board computer for a robot, one has to consider the ways of use and needs of that particular robot. For example, the HTC Google Nexus One can be found unlocked for less than $200, and is a suitable on-board computer.

- **Software**

The Android operating system is open source and Linux-based. Programmers can develop software for Android in Java using the SDK [4–6] or in native language (C/C++) using the native development kit (NDK). It is also possible for developers to modify the Linux kernel if needed. Implementation of an Android application can be achieved using the Eclipse IDE with the Android Development Tools (ADT) plug-in.

Using this SDK, the developer has easy access to different functionalities of an Android phone such as graphical interfaces, multi-threading, networking, data storage, multimedia, sensors, location provider, speech-to-text, text-to-speech and more [6].

3 Course of designing game

Game design is the art of applying design and aesthetics to create a game for entertainment or for educational, exercise, and experimental purposes. Increasingly, elements and principles of game design are also applied to other interactions, particularly virtual ones.

Academically, game design is a part of game studies, while game theory studies strategic decision making (primarily in non-game situations). Games have historically inspired seminal research in the fields of probability, artificial intelligence, economics, and optimization theory. Applying game design to itself is a current research topic in meta-design [7].

3.1 Game development

To develop games for Android, you need a PC running a version of either Linux, Windows, or Mac OS that meets the requirements for both the Android SDK and the Java Development (JDK). Both the SDK and JDK are freely available from their respective websites, where you can find more details about specific requirement. Android also uses the Eclipse IDE (integrated development environment).

Working knowledge of Java and XML are helpful, but not necessary. If you are familiar with any high-level language and development environment, you should be fine. If not, you should still be able to work through the examples and put together workable games.

If you are interested in developing for Android, you probably have an Android device, but you do not necessarily need one. The Android SDK provides an emulator, which lets you configure virtual devices to test your games without the actual hardware.

However, testing playability without actual devices is not advised. Especially if you are designing for multiple form factors, such as both phones and tablets, you will probably want to invest in at least a couple of test devices.

3.2 Game architecture

Games have a different architecture and control flow than apps. Both seem to respond to user input instantly, but while an app does this by setting listeners and reacting to events with method call (most commonly the onClick method calls the OnClickListener), this approach is not valid for a real-time game (although it is valid for non-real-time games).

Once a game is running, it must evaluate and update everything as fast as possible. This is the reason why it cannot be interrupted by user events. Those events or states should be recorded instead and then read by the game objects during its update (Figure 3) [8].

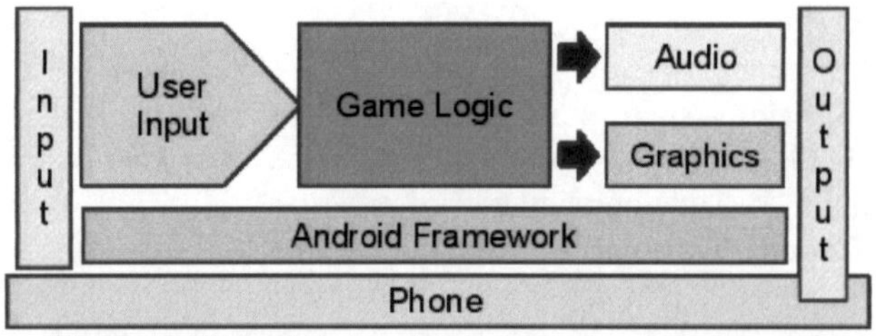

Figure 3. Simplified architecture of a game engine.

4 Robotics in Lao

Engineering students compete for robotics prize. According to the National University of Lao, the faculty of engineering is encouraging students to gain science and technology experience by giving them the opportunity to design and build robots so that they can demonstrate their achievements in the field of robotics.

Currently, in Lao PDR, the reported and promoted information in the field of science and technology is not enough. Therefore, we have the goal of developing a web and mobile portal in order to provide Lao people with an opportunity to get access to information and resources that can be useful and easily reached [9–11]. In addition, science and technology information and resources will enable integration of Lao people to the scientific society.

5 Conclusion

The paper presents a promising trend in robotics, which leverages smartphone technology. These smartphone robots are ideal for hobbyists, educators, students and searchers. The paper describes smartphone-based robotic projects and demonstrates the relatively easy and inexpensive construction of an Android-based robotic platform.

Acknowledgments

This study (research grant No № 8.2.24.2018) was supported by the Tomsk State University competitiveness improvement programme.

References

[1] Android intro. [Electronic resource] URL: https://www.elprocus.com/what-is-android-introduction-features-applications/
[2] Related work. [Electronic resource] URL: http://journals.sagepub.com/doi/full/10.1177/1729881417716010#_i11
[3] Humanoid robot. [Electronic resource] URL: https://en.wikipedia.org/wiki/Humanoid_robot
[4] ERICA. An autonomous conversational robot. [Electronic resource] URL: https://geminoid.jp/en/robots.html
[5] Android based robotic platform. [Electronic resource] URL: http://socsci.uci.edu/~jkrichma/ABR/
[6] Android software and hardware. [Electronic resource] URL: https://apress.com/gp/book/9781430244349
[7] Designing game. [Electronic resource] URL: https://en.wikipedia.org/wiki/Game_design
[8] Game architecture. [Electronic resource] URL: https://packtpub.com/mapt/book/game_development/9781787128583/21/ch01lvl1sec00010/game-architecture
[9] Solinthone Ph, Rumyantseva T 2016 *MATEC Web of Conferences* 7 01066 doi: 10.1051/matecconf/20167901066
[10] Solinthone Ph, Rumyantseva T, Phongtraychack A 2018 *MATEC Web of Conferences* **155** 01034 doi:10.1051/matecconf/201815501034

[11] Serikov M Z *et al* 2017 *Proceedings - 2016 11th International Forum on Strategic Technology, IFOST 2016* 510

Clinical trials of a personal electrocardiograph

A Boyakhchyan[1], I Lezhnina[1], K Overchuk[1], V Perchatkin[2], A Lvova[3] and U Alexander [4]
[1]Tomsk Polytechnic University, Tomsk, Russia
[2]Head of the Department of Rehabilitation of Patients with Cardiovascular Diseases, Cardiology Research Institute, Tomsk, Russia
[3]Postgraduate Student, Cardiology Research Institute, Tomsk, Russia
[4]Project manager, DI-Group, Moscow, Russia
E-mail : inna84-08@mail.ru

Abstract. The article describes the results of a clinical trial at the Cardiology Research Institute in Tomsk. Clinical trials were conducted to identify different important information for diagnosis. These tests were also conducted to remotely monitor the treatment of patients who had already been discharged from the hospital.The study involved 15 patients, the most interesting cases are described in this article.

1 Introduction
At present electrocardiography allows to diagnose and control many conditions and pathologies of the heart: various rhythm/conduction disorders, control the therapy of certain medications, etc [1].

Electrocardiographs are a complex device that should be managed by a specialist who has special skills and is able to correctly register an ECG, which absolutely excludes the possibility of using this equipment outside the cardiologist's office.

To solve this problem, developed a personal electrocardiograph.To verify the appropriateness of using such a device, were conducted clinical trials [2–6].

Research s plan:

- Tests are carried out by practical application of ECG-Express devices on patients with selected groups of heart diseases.
- Patients use devices to self-register the ECG during the onset of symptoms, and also on a regular basis as directed by doctor.
- For the measurement, different ECG leads are selected depending on the location of the expected signs in different leads: according to the standard arrangement of the device in the center of the chest, and in other positions [6–12].

2 Materials and methods
During the testing period, the device was given to 15 patients for a period of one to two weeks for regular using. A common feature for each individual case was that at previous diagnostic stages it was not possible to clarify the diagnosis. Patients had rare complaints about the heart, and with daily monitoring of ECG Holter pathology could not be registered.

The study group of patients included:

- 3 patients with IHD (ST segment ECG inversion);
- 2 patients with myocarditis (inversion of the T wave);

- 2 patients who control the dose of antiarrhythmic drugs;
- 8 people with various heart rhythm disturbances.

For registration of ECG, was used a personal electrocardiograph "ECG-Express" – development of scientists from Tomsk Polytechnic University (Figure 1).

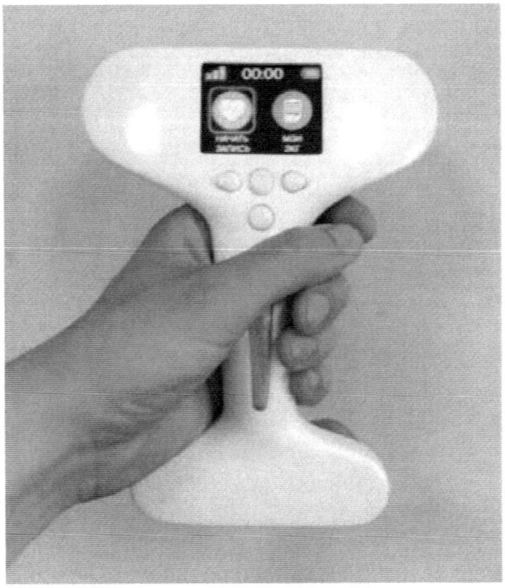

Figure 1. The device "ECG-Express".

The device satisfies a number of basic requirements set out in the development of:

- Device is personal and mobile so that the patient can use it in any life situation.
- Measurement occurs from the surface of the chest, because it chest leads registration provides important information for the diagnosis.
- Measurement is equally qualitative and stable in people with different individual characteristics: chest anatomy, hair density, contamination and skin condition.

The ECG is recorded for 50 seconds, with 3 chest leads. By means of mobile communication, information is sent to the server from which the doctor in the specialized application performs an ECG analysis. All records are stored on the server and computer of the doctor in chronological order.This device was offered to patients who were discharged from the hospital and continued treatment on an outpatient basis. Patients independently used it at home.

3 Results

During the testing of the device, a number of special cases were singled out, when the application of the device made it possible to identify symptoms that could not be fixed by standard methods. The following is a description of the main, most significant, cases.

In other cases, using a device able to confirm the diagnosis made in a standard way of ECG registration. All ECG was opened and filtered in MatLab. Screenshots with filtered signals are presented below. MatLab has a Russian interface and all screenshots below from MatLab.

The patient is a man, 42 years old. He was at the Research Institute of Cardiology on examination and treatment with a diagnosis: IHD. On an outpatient basis he complained of pain in the chest like pressing. Patient independently used a personal electrocardiograph (Figure 2), thanks to which it was possible to register changes in the T wave. He was hospitalized, was revealed postmiocardic cardiosclerosis.

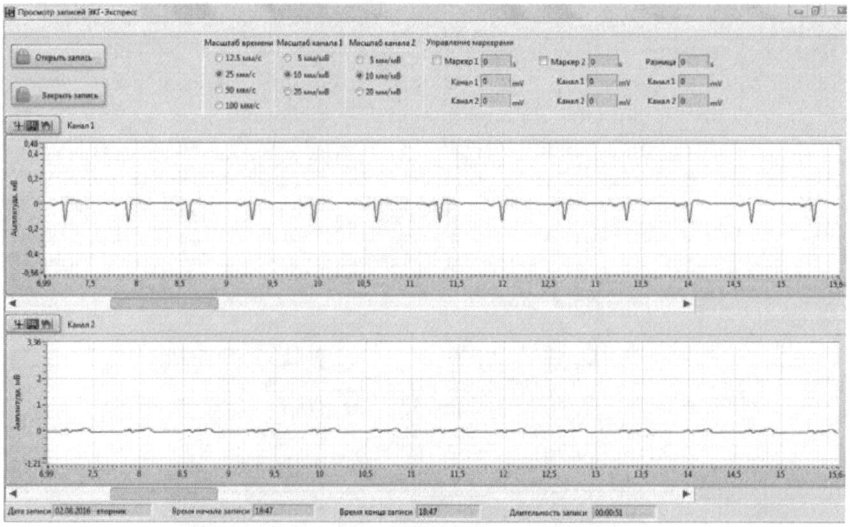

Figure 2. Screen of ECG of the first patient.

The patient is a man, 78 years old with a diagnosis: ischemic heart disease and paroxysmal ventricular tachycardia. This patient is stable heavy as it has damage to the spine, and therefore is always at home. Bthe possibility to visit outpatient clinics and to be constantly monitored by a doctor is not available for patient. Suddenly, in this patient there was deterioration. The use of a personal electrocardiograph made it possible to diagnose the onset of paroxysm of atrial fibrillation and in time to call for an ambulance, without waiting for the further development of negative symptoms.

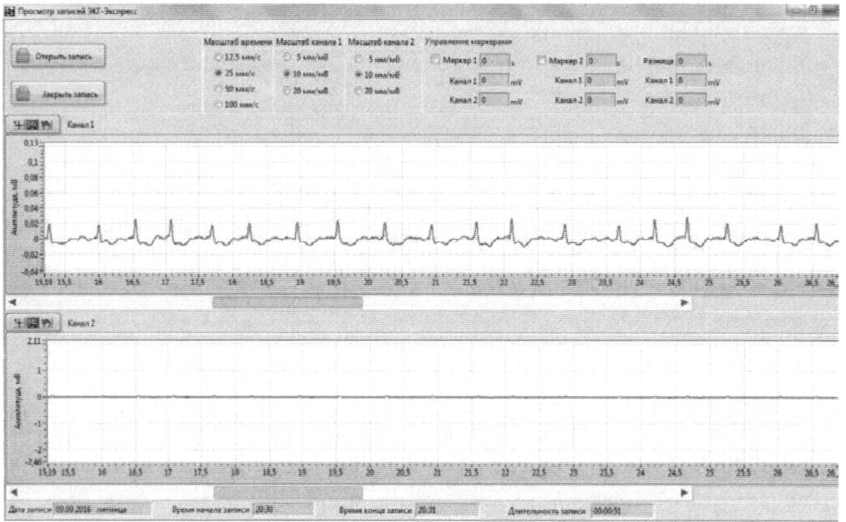

Figure 3. Screen of ECG of a patient with ischemic heart disease.

The patient is a young man, 17 years old. The young man complained of sudden attacks of palpitations. Repeated daily monitoring of ECG pathology did not reveal. With the help of a personal electrocardiograph (Figure 3), it was possible to diagnose paroxysms of supraventricular tachycardia. Later the patient was treated in the arrhythmology department.

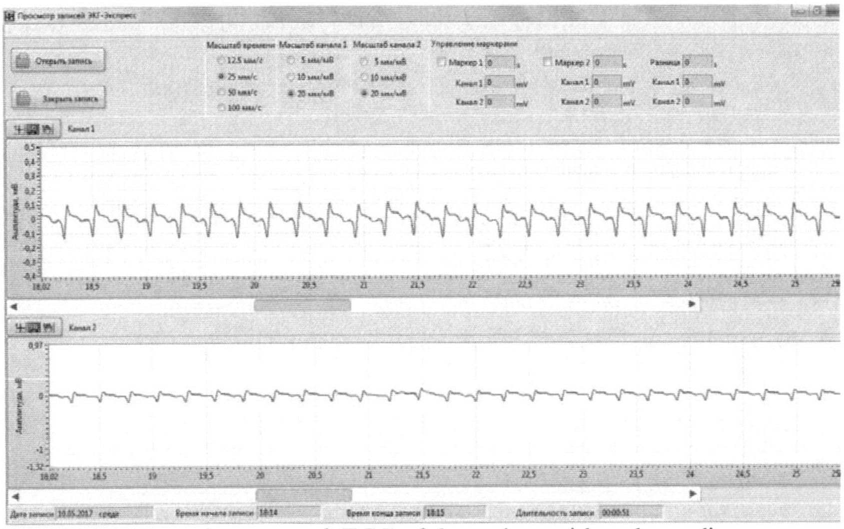

Figure 4. Screen of ECG of the patient with tachycardia.

Patient is a man of 52 years old. Was observed outpatient with the therapist at the place of residence. Was consulted by one of the staff of the Institute of Cardiology. He complained of fits of heart failure, accompanied by mild dyspnea, weakness, suddenly started, lasted 3–5 minutes and self-stopped. These episodes arose several times a month. Conducting daily monitoring of the ECG in this case was not informative. The patient used a personal electrocardiograph (Figure 4) while in another city. It was possible to establish the probable cause of the described seizures and record a rhythm disorder - episodes of frequent ventricular extrasystole. After the examination in the cardiological hospital, the following diagnosis was made: Chronic nonspecific focal myocarditis.

4 Conclusion

Analyzing the obtained data, it can be concluded that a personal electrocardiograph can be used for:

1. Detection of irregular heart rhythm disturbances, which can not always be detected with the help of 24-hour ECG monitoring.
2. Receiving valuable information for diagnosis (ECG recording during an attack).
3. To establish possible episodes of myocardial ischemia.

When registering the ECG, patients did not notice any difficulties in using the device. The ability to transfer a record immediately to the doctor's workplace allows you to react immediately, and take appropriate measures, while the patient can be thousands of miles away from clinic.

References

[1] Montalescot G., Sechtem U., Achenbach S., Andreotti F *et al.* 2014 *European Heart Journal* **35 (33)** 2260–2261 doi :10.1093/eurheartj/ehu038

[2] Lezhnina I A, Boyakhchyan A A, Overchuk K V and Uvarov A A 2017 *IOP Conf. Ser.: Mater. Sci. Eng.* **881** 012041 doi : 10.1088/1742-6596/881/1/012041

[3] Avdeeva D K, Kazakov V Y, Natalinova N M, Maksimov I V, Balahonova M V 2014 *Biology and Medicine* **6(2)** BM-025

[4] Bordunov S V 2016 *IOP Conference Series: Materials Science and Engineering* **110** 012094 doi: 10.1088/1757-899X/110/1/012094

[5] Kazakov V Y, Avdeeva D K, Grigoriev M G, Natalinova N M, Maksimov I V, Balahonova M V 2015 *Biol Med (Aligarh)* **7(5)** BM-155-15

[6] Natalinova N, Ilina N, Frantcuzskaia E 2016 *IOP Conf. Ser.: Mater. Sci. Eng* **132** 012029

[7] Montalescot G, Sechtem U, Achenbach S, Andreotti F *et al.* 2013 *European Heart Journal* **34** **(38)** 2949–3003 doi : 10.1093/eurheartj/eht296
[8] Golitsyn S, Kropacheva E and Maikov E et al. 2013 *Federal clinical guidelines for the diagnosis and treatment of cardiac arrhythmias and conduction: wedge.* (Moscow , Recommendations)
[9] Komolyatova V, Kupriyanov O, Pervova E and Ryabykina G 2014 *Russian Cardiology Journal* **(2)** 6–71 doi : 10.15829/1560-4071-2014-2-6-71
[10] Yartsev S 2014 *Electrocardiography: a practical guide-guide for doctors* (Moscow, RUDN)
[11] Plotnikova I V *et al* 2018 *IOP Conference Series: Materials Science and Engineering* **289** 012029 doi:10.1088/1757-899X/289/1/012029
[12] Axelrod A, Chomakhidze P and Syrkin A 2007 *Holter monitoring of ECG: possibilities, difficulties, errors* (Moscow, Medical Information Agency LLC)

The study of features of the structural organization of the automated information processing system of the collective type

V N Nikolaev[1], D V Titov[1], V I Syryamkin[2]
[1] The Southwest State University, Kursk, Russia
[2] National Research Tomsk State University, Tomsk, Russia
E-mail: nikovic54@yandex.ru

Abstract. The comparative assessment of the level of channel capacity of different variants of the structural organization of the automated information processing systems is made. The information processing time assessment model depending on the type of standard elements and their structural organization is developed.

1 Introduction

Nowadays the Use of informational technologies and the automated information processing systems (AIPS) in various spheres of economy and the industry showed their high efficiency at the solution of a wide class of tasks [1–4]. At the same time the analysis and study of the processes of assessment and comparison of various forms of the structural organization of the AIPS on the basis of standard elements and different types of program information support is important and critical [5, 6].

2 The theoretical part

In a number of scientific sources the variants of the organization of the automated information processing systems (AIPS) of various functional purposes were considered. In the applied theory of the mass service the mathematical models [5, 6] allowing to receive the quantitative assessments of a number of parameters of functioning of the systems of such class are developed. In the same place it is shown that there are different approaches to the organization of polyphase subsystems of AIPS that allows to combine the program, technical means, information and intellectual support depending on the tasks solved by AIPS, and requirements to the systemic resources. Therefore for the purpose of obtaining recommendations about the use of rational variants of creation of the AIPS with different structure and characteristics of elements we will make the quantitative assessment of their effectiveness in the range of the systemic parameters which are of practical interest.

We will choose the variants of creation of AIPS on the basis of the decentralized and centralized structures as the studied ones (Figure 1).

The presented variants are the main types of structures allowing to create the complex multipurpose AIPS [7, 8]. The mathematical models of processes of their functioning are rather well developed. As the measure of comparison we use the efficiency factor [1] estimating the average intensity of an output informational stream from the standard element of the AIPS. The ratios of the indicators characterizing the superiority on performance level of one variant of creation of the AIPS in comparison with another one will be the criterion of efficiency.

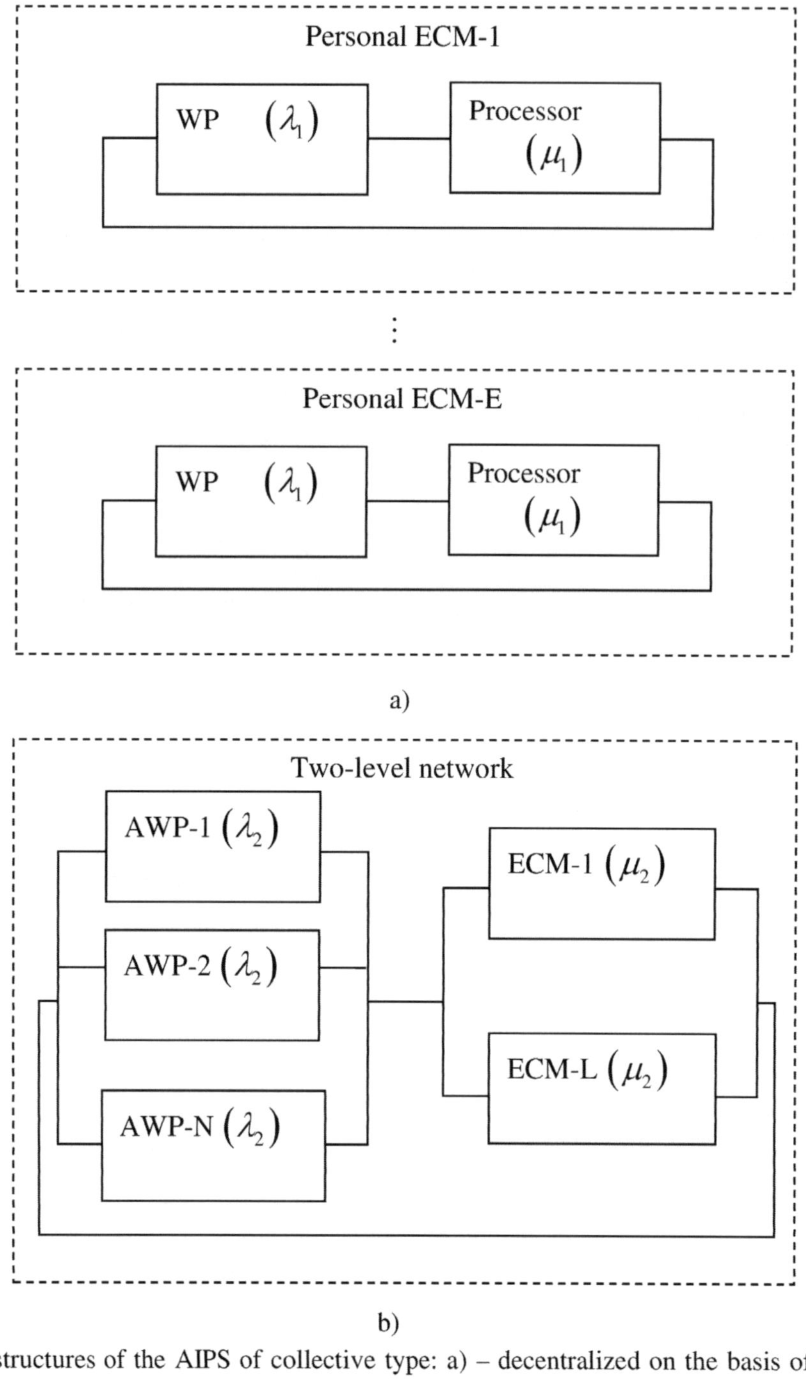

Figure 1. Basic structures of the AIPS of collective type: a) – decentralized on the basis of personal computers; b) – centralized on the basis of two-level network.

With the purpose of study of influence of different structural features of creation of subsystems of the AIPS we will accept the following restrictions and assumptions.

The total intensity of processes of transformation of information at each phase of the system is identical in sense of the following ratios [5, 6]:

$$\sum_{i\in I}\lambda_i' = \sum_{j\in J}\lambda_j'' = \lambda; \qquad (1)$$

$$\sum_{l\in L}\mu_l' = \sum_{s\in S}\mu_s'' = \mu; \qquad (2)$$

$$\rho = \lambda/\mu,$$

where *I, J* и *L, S* – respectively the number of elements of the competing variants at each level of creation of the system.

At the same time we will accept that the costs of realization of the variants are also equal. Such assumption generally is not true, but it allows to simplify the analysis and to make the quantitative assessment of the influence of structure and characteristics of the variants on channel capacity of the AIPS.

Taking into account the above-said we will take the loss in channel capacity of one variant in relation to another as the measure of comparison [7, 8]:

$$\omega(B) = 1 - B'/B'' \qquad (3)$$

or

$$\omega(B) = 1 - \frac{E\mu'\overline{L_a'}}{\mu''\overline{L_a''}},$$

where μ', μ'' – respectively the intensity of service of applications by the information-computer environment; $\overline{L_a'}$, $\overline{L_a''}$ – respectively the mean number of the active computers (servers) in the compared variants; E – the number of identical structures in the variant.

The schedules of dependence of the indicator (3) as a function from the parameters *E, L, N*, ρ are presented at the Figure 2.

As their analysis shows [9,10], at observing the conditions (1,2) and with any values of *E, L, N*, ρ the centralized structure realized in the form of the collective AIPS in comparison with the variant on the basis of independent personal computers is more preferable on the level of a channel capacity. The difference is the highest in the range of parameters $10^{-1} \le \rho \le 10$, $1 \le l \le 8$, $8 \le N \le 50$. Here the variant with the centralized structure has the channel capacity by 10÷45 % higher than with the decentralized one. The maximal difference is at the values of the parameters $L = 1$, $N > 50$ and $\rho = 1$-2 and reaches the value of 50%. At $L = N$, and also in the range of $\rho < 10^{-2}$ and $\rho > 10^{2}$ essential differences are not observed and the variants are equivalent in the channel capacity. In this case it is necessary to consider other indicators entering the criterion of effectiveness (3).

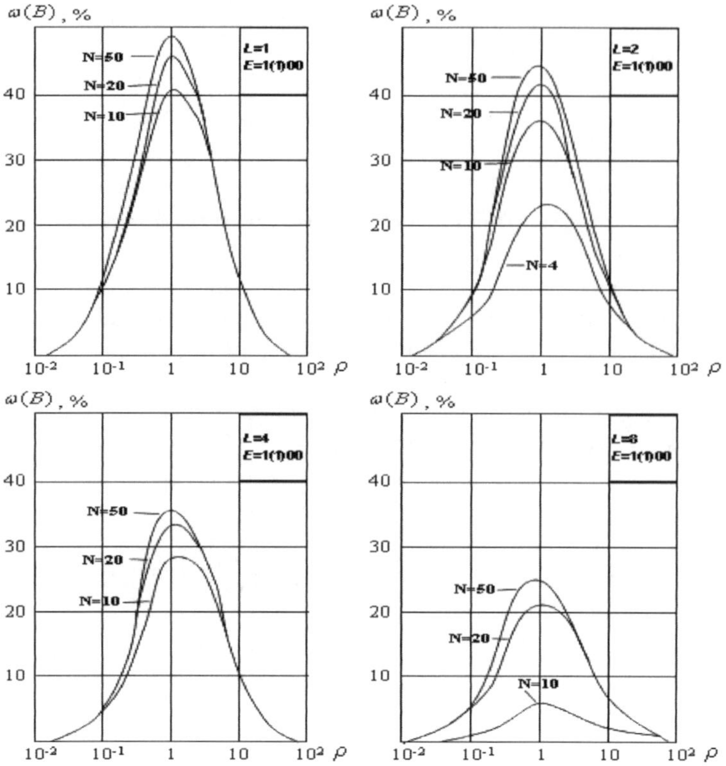

Figure 2. The comparison of the level of channel capacity of the variants of creation of AIPS: the decentralized one on the basis of personal computers of the type $E\{M/M/1(\mu')/1(\lambda')\}$and the centralized one on the basis of 2-level network.

It is also necessary to notice that the change of the indicator E for the decentralized variant does not influence the value of the expression (3). It is connected with the fact that the increase in number of personal computers linearly increases the channel capacity of the whole system, but at the observance of the conditions (1, 2) the efficiency of separate structure also linearly decreases.

The need of use of hierarchical two-level structure conforms well to the principles of division of tasks and resources.

The best efficiency of the centralized structure is reached due to higher value of the coefficient of use of computing resources. However the centralization of computing tools at $N > L$ leads to emergence of turns at the requirement of systemic resources therefore the time of processing of the array of information increases.

Therefore with the purpose of the study of influence of parameters of the considered variants on the time of transformation of information we will introduce the function [11]

$$\omega(T_{np}) = \overline{T''_{np}} \Big/ \overline{T'_{np}}, \qquad (4)$$

where $\overline{T''_{np}}$, $\overline{T'_{np}}$ – the mean time of transformation of the array of information in the centralized AIPS on the basis of network and the decentralized structure on the basis of personal computers respectively.

Using the similar approach, we will present the expression (4) as

$$\omega(T_{np}) = \frac{N\mu'\overline{L'_a}}{E\mu''\overline{L''_a}}, \qquad (5)$$

where E – the number of personal computers; N – the quantity of ARM in the network.

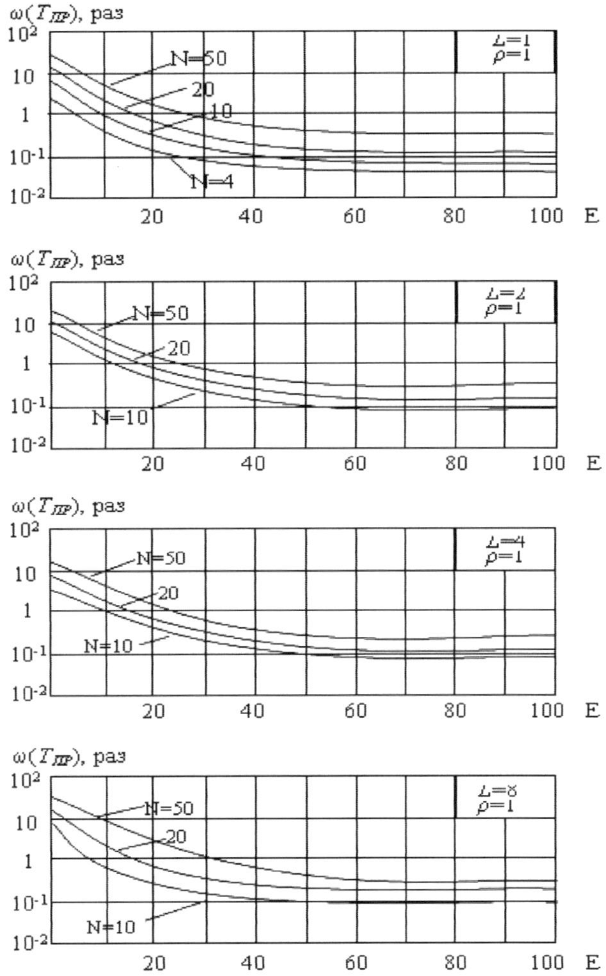

Figure 3. The comparison of the time of transformation of information for different variants of creation of AIPS.

The analysis of results of the study of (5) as the function from the systemic parameters E, L, N, presented in the Figure 3 shows that time of transformation of information strongly depends on the values of the listed parameters.

TSo, at the observance of conditions $N \geq E$ and with any L, N, ρ the decentralized structure on the basis of personal computers will be more preferable in sense of the time of transformation of information, i.e. the inequation $\omega(T_{np}) > 1$ will be true.

At $N < E$ and defined values of the parameters L, N the information processing time in the centralized structure can be less, than in the decentralized one.

The increase in the parameter N leads to the increase in the of values of the function (5), and for the parameter L the inverse dependance is observed, i.e. the higher is the coefficient of centralization of computing means N/L, the less preferable is the network variant of creation of AIPS, than the variant on the basis of personal computers other factors being equal.

In general the duration of transformation of information for the considered variants in the range of parameters $E = 1(1)100$, $L = 1(1)8$, $N = 1(1)50$, $10^{-2} < \rho < 10^{2}$ can differ by more than two orders.

The made analysis allowed to reveal the regularities and to receive the quantitative assessments of temporary parameters of functioning of the AIPS depending on its structure and characteristics of the making elements.

3 Conclusion

The made analysis allowed to reveal the regularities and to receive the quantitative assessments of temporary parameters of functioning of the AIPS depending on its structure and characteristics of the making elements.

Acknowledgments

This study (research grant No № 8.2.24.2018) was supported by the Tomsk State University competitiveness improvement programme

References

[1] Nikolaev V N 2015 *Technical and economic models of geographic information systems of the scientific and production enterprises: monograph* (Kursk, Publishing house of SSU)
[2] Nikolaev V N 2014 *The methodology of construction and use of innovative geographic information systems: monograph* (Kursk, Publishing house of SSU)
[3] Nikolaev V N 2013 *Management of life cycle of innovative GIS: monograph* (Saarbrücken, Palmarium academic publishing)
[4] Nikolaev V N, Dafalla A A B, Pleskanos L V, Kassim K D A 2013 *Telecommunications and Radio Engineering* **72(13)** 1219–1225
[5] Kalashnikov B B, Rachev S T 1988 *Mathematical methods of creation of stochastic systems of models of services* (Moscow, Science)
[6] Buslenko N P 1968 *Modelling of the complex systems* (Moscow, Science)
[7] Nikolaev V N 2005 *Control system and informational technologies* **(3)** 70–74
[8] Nikolaev V N 2006 *Control systems and informational technologies* **3(25)** 19–22
[9] Maslak A A, Nikolaev V N 2013 *Information and measuring and operating systems* **11(8)** 46–49
[10] Yurchenko A *et al* 2016 *MATEC Web Conf.* **79** 01085
[11] Makariun I V, Nikolaev V N 2013 *Proceedings of higher educational institutions. Geodesy and aerial photography* (**1**) 93–100

Advanced composite alloys for constructional parts of robots

D K Issin[1], Zh D Zholdubayeva[1], Y G Neshina[2,3], A D Alkina[3], N Khuangan[1] and G M Rahimova[1]
[1]Karaganda state technical university, Karaganda, Kazakhstan
[2]Tomsk State University, Tomsk, Russia
[3]Tomsk Polytechnic University, Tomsk, Russia
E-mail: 1_neg@mail.ru

Abstract. In recent years all over the world special attention has been paid to the development and implementation of nanostructured materials possessing unique properties and opening fascinating prospects for the development of technical progress in various fields of human activities. A special place can be given to the development of service robots, the market of which is actively developing. There is problem associated mainly with the lack of heat-strengthened alloys which consists in low thermal stability of the alloy properties under the conditions of elevated variable temperatures and loads. The article presents studies to assess the effect of composition, the amounts of refractory nanoscale particles and methods for their introduction into the melt on the structure and properties in nanostructured composite aluminum alloys. The powders of metals, alloys, as well as silicon carbide and aluminum oxide were used to produce the nanostructured powder composite materials. As a result of the research, NPCM compositions containing micro-size particles of transition metals that are carriers of nanosized reinforcing particles and initiators of the formation of an intermetallide of endogenous origin in a melt.

1. Introduction

The development of robotics creates the significant prospects for replacing the human in hazardous and difficult working conditions in the industry and will lead to the replacement of a number of professions that will eventually become history. Today there are the production facilities with a high level of automation where about two or three dozen people employs, and the rest of the staff is replaced by industrial robots and this is not uncommon. According to various estimates of experts in the field of robotics the level of production of industrial robots ranges from 20 to 30 billion dollars. While a smaller amount of them is allocated for service robotics. The main buyers and manufacturers of industrial robots are USA, Germany, Japan, South Korea, China. The world leader in production remains Japan, which owns 50% of the world market, about 20% in Germany, but in connection with the entry of China into the world market, the situation may change in the coming decades.

The development of robotics is a catalyst for the development of new industry 4.0. A special place can be given to the development of service robots, the market of which is also actively developing due to the enormous volume of various developments that will be so accessible in the next 10 years that every family in half the countries of the world will can afford to buy a service robot. These development prospects form the specific problems, in particular, one of which is the search and creation of new materials for the manufacture of constructional parts of robots. Materials must have a number of properties that allow them to be used in mass production at a low cost, but it should not be at the expense of their mechanical strength. The strength of the materials and their availability will make it possible to expand the range of their applicability and availability. There is no doubt that not only the development of materials of the new generation is essential, but it will be necessary to promote science in the field of nanotechnology, precision mechanics, electronics and optics. In our work, for further processing and production of constructional parts of robots we would like to touch on the theme of finding new composite materials based on aluminium matrix alloys, consider the methods of their casting and the melt of the particles of the reinforcing filler (short fibers, whiskers, particles), and the subsequent filling of the resulting liquid-metal suspension into a casting mold after cooling of which the required product is obtained.

2. Analysis of the problem

There is a small problem associated mainly with the lack of heat-strengthened alloys which consists in low thermal stability of the alloy properties under the conditions of elevated variable temperatures and loads. Therefore, the developments in the field of synthesis of composite aluminium matrix alloys (CAMA) are started and being actively implemented, the main feature of which compared with known ones, including high-strength alloys is the presence in the structure of artificially introduced or initiated as a result of exothermic reactions thermos table components that provide the required properties of the material, including at elevated (up to 200°C) and high (up to 400 °C) temperatures. The properties of CAMA are determined by the properties of the matrix material, properties and dimensions of the reinforcing ones introduced or initiated (obtained in-situ) into the melt of particles and the interaction between the matrix and the reinforcing particles. Ready-made micro- and nanoscale refractory ceramic particles (for example, SiC, Al_2O_3, Si_3N_4, AlN, B_4C, MgO), intermetallic compounds and other types of particles can be used as reinforcing complexes introduced into CAMA. Fundamentally important thing in obtaining CAMA is to ensure the wettability of the introduced particles with a matrix melt to form the stable adhesion bonds.

Three technological schemes of CAMA production are mainly used: introduction of particles into the melt under intensive mixing with the help of an impeller; powder technology; impregnation of dispersed particles or press molds with a matrix melt. Liquid-phase technology for the production of KAMA with the introduction of a significant amount (up to 20%) of micrometric refractory particles into the melt has been achieved to widespread industrial applicationы (MS 21, Duralcan, Science Application Intern. Corp. (USA), British Alcan, Talon Composites (England) and etc.). A number of companies, in particular Duralcan, supply the resulting alloys in the form of cast sections (billets), which are subsequently used as the charge materials in obtaining of cast products by various methods of casting. Depending on the composition of the matrix alloy, the composition and the amount of particles introduced, alloys with different complex of mechanical and operational properties are obtained, used for products of different purposes.

3. Materials and methods

The method of composite casting (Compocasting) includes mixing of reinforcing filler particles (short fibers, whiskers, particles) into the wide-interval matrix melt with a simultaneousgradual decrease in temperature and then pouring the obtained liquid metal suspension into the casting mold after cooling of which the required product is obtained. The method is developed on the basis of rheological casting (Rheocasting) created in the early 1970s by Fleming and others. The main advantage of the method is the possibility of introducing the non-wetting particles that are mechanically captured by the melt.

In recent years all over the world special attention has been paid to the development and implementation of nanostructured materials possessing unique properties and opening fascinating prospects for the development of technical progress in various fields of human activities. Despite a significant number of works of foreign researchers in which the attempts are made to theoretically substantiate the processes occurring in nanostructured composite aluminum alloys (NCAA), the theory of these processes has not been created. There are only separate attempts to explain the phenomena that occur in the NCAA in the process of forming their structure and properties.

The problem of developing nanostructured metal matrix compositional alloys is studied by many researchers in the world largest scientific centers, as well as in a number of Russian research and educational institutions. After analyzing the scientific papers [1–7], it can be noted that only a part of them is devoted to the development of casting composites, i.e. alloys with a level of casting properties, which allow obtaining from them the shaped cast products. In these works, the main requirement for such alloys and components is a certain level of tribological properties. The main methods of preparation of dispersed particles is the mechanical alloying, the main method for introduction into the melt is the mechanical mixing of compressed tablets or briquettes. The importance and effectiveness of the development of NCAA is confirmed particularly by the attention given to this problem abroad. Based on the results of works that was performed by us [8–11] on obtaining the NCAA on alloy A357basis, at SPbSTU*) jointly with Zvezda OJSC**) the studies have been conducted to assess the impact of composition, the amount of nanoscale refractory particles, and the methods of their introduction into the melt on the structure and properties of NCAA. One of the most serious problems in obtaining NCAA is the preparation of nanoparticles for introduction into the melt. In most cases, the nanostructured powder composite materials (NPCM) obtained by mechanical alloying consisting of particles of a metal carrier powder with nano-dispersed non-metallic inclusions distributed therein are used. It has been theoretically justified and experimentally proven that the properties of composites are increased with a decrease in dispersion of reinforcing phases to the nanometer level.

For the production of NPCM we used the advanced technology of activation mechanical alloying in high-energy mills (attritors) [8]. The powders of metals, alloys, as well as silicon carbide and aluminum oxide were used to produce NPCM.

NPCM (Cu + 4.5% Al_2O_3) was obtained by thermochemical method [9, 12] using aqueous solutions of copper and aluminum nitrates (Figure 1). This technology ensures the uniform distribution of nanoscale refractory particles in a micrometric basis directly during the synthesis of NPCM.

NPCM can be seen not only as a raw product in the production of NCAA, but also as an independent commodity product, which can be supplied to various manufacturers of products from composite metal-matrix alloys.

The composition of the parent alloy A357 (mass. %): 7.5 Si, 0.35Fe, 0.45 Mg, 0.02 Mn, 0.06 Ni, 0.03 Cu, 0.08 Zn, Al – the rest. The properties of as-cast parent alloy A357:$\sigma_{ultimate}$=155 MPa, δ = 3.5 %. In the first series of experiments, A357alloy (2 kg) was heated in the resistance furnace to 800 °C, the calcined all-purpose flux was mixed and removed, the crucible was transferred to the electromagnetic stirrer, 5% of NPCM (62.5% Cu + 28.1% Ni + 3.1% SiC) was introduced during the mixing with a plasma facility.

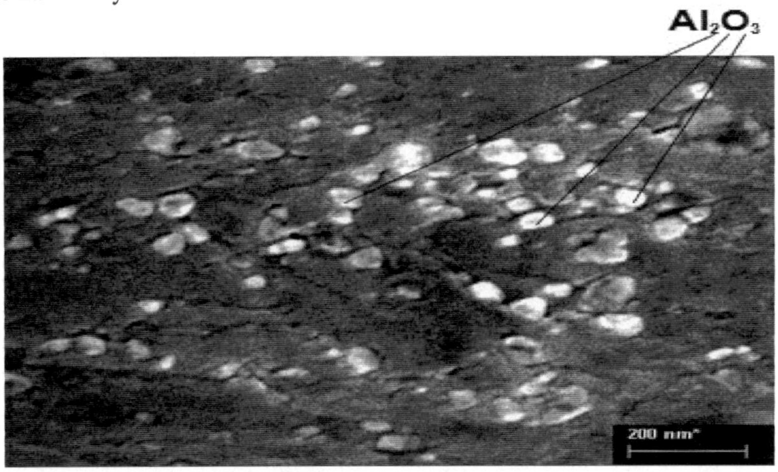

Figure 1. Structure of the NPCM Cu/Al2O3 (thermochemical method).

The duration of the NPCM introduction (work of plasma facility and stirrer) is 50...60 s, the temperature of the metal in the process of NPCM introduction dropped to 720...750°C after which the casting mold was filled with samples with a diameter of 12 mm and a length of 100 mm. The NPCM calculated assimilation was 94% before heat treatment at the normal temperature $\sigma_{ultimate}$=280 MPa, δ = 3.5 per cent. After T6 at the normal temperature $\sigma_{ultimate}$= 390 MPa, δ = 7.5 %, at 350°C-$\sigma_{ultimate}$=160 MPa, δ = 11%. Metallographic studies of the obtained A357 alloy (SiC) confirmed the presence of nanoscale SiC inclusions, the invariance of the phase composition and the dimensionality of these inclusions during the NCAA synthesis. In the second series of experiments, the conditions were identical to the first series, however, 5% of the NPCM containing 95.5% Cu + 4.5% Al_2O_3 was introduced into the melt with a plasma facility. The NPCM calculated assimilation was 92%, before the heat treatment at the normal temperature$\sigma_{ultimate}$= 290 MPa, δ = 4.5%. After T6 at the normal temperature, $\sigma_{ultimate}$ = 405 MPa, δ = 8.5%, at 350 ° C - $\sigma_{ultimate}$ = 165 MPa, δ = 12.5%. After T6 at the normal temperature $\sigma_{ultimate}$ =405 MPa, δ = 8.5%, at 350°C-$\sigma_{ultimat}$=165 MPa, δ = 12.5%.

In the course of preliminary experiments, the method of introduction into the melt of NPCM with the help of the "flux-cored wire" variant (introduction of an aluminum tube filled with NPCM under the bath level in furnace) was also used when mixing NPCM into the melt with a mechanical stirrer (300 rpm) and then transferring the crucible with melt into the electromagnetic stirrer. The temperature regimes, composition and number of NPCM are identical to those realized in plasma synthesis. The NPCM calculated assimilation was 76%, before the heat treatment at the normal temperature, $\sigma_{ultimate}$= 205 MPa, δ = 3.5%, after T6 - 280 MPa, δ = 4.0%.

The plastic treatment of the resulting cylindrical cast sections (billets) in a liquid-solid state (585 ± 5 °C) confirmed the possibility of practically eliminating the initial porosity, realizing the process of liquid-solid molding (thixocasting) with the invariable distribution of nanoscale particles of the reinforcing phase.

4. Summary

It is possible to introduce the nanoscale reinforcing and modifying particles into the melt of NPCM, it can be a separate type of commodity product. The most promising are the compositions of NPCM containing micro-sized particles of transition metals that are carriers of nanoscale reinforcing particles and initiators of formation of intermetallicsof endogenous origin into the melt (in-situ process).The behavior of nanoparticles in the melt (agglomeration, dissolution, uniformity of distribution) during the introduction and mixing of particles, exposure, during the filling of the composite into the mold and the formation of the casting product, the influence of the amount of phase composition, the conditions of introduction of nanoparticles into the melt on the properties of composites are virtually absent or debatable. The use of wide-interval alloys as a basis for the production of NCAA allows to intensify the effect of wettability of particles by melt during its mixing in the liquid-solid state, ensure the uniformity of their distribution in the volume of melt and ingot (casting), use the melt in the liquid-solid state to produce the dense products by thixocasting method. Two-plane electromagnetic stirring of melt, regardless of the technology of nanoparticle introduction into the melt and the pre-mixing process, provides the possibility of mixing the melt in both liquid and liquid-solid states. The process of plasma injection of NPCM into metal melts in combination with two-plane electromagnetic stirring has a number of advantages over other methods of obtaining NKAA, but requires optimization of operating modes depending on the type of matrix alloy, the composition and number of introduced particles, the volume of production of billets and shaped products, and also requires the feasibility study of advantages over other methods of NKAA obtaining in specific production situations. In order to obtain the NCAA with improved performance properties, it is necessary to realize all the technological possibilities for improving the properties of the matrix alloy, in particular, nanomodification and nanoalloying, external influences on the liquid and crystallizing alloy. The nanostructural composite metal matrix alloys are one of the most promising construction materials of the future and their use in robotics. The liquid-phase technologies for their production are the most competitive in the production of products from the NCAA on an industrial scale in order to reduce the cost of constructional parts of the mechanical robots.

Acknowledgments
This study (research grant No № 8.2.24.2018) was supported by the Tomsk State University competitiveness improvement programmer.

References

[1] Petrunin A 2009 *Development and mastering of the technology of obtaining alumo-matrix composite alloys modified with nanoparticles.* Author'sabstract. diss. Ph.D. VSTU, Vladimir
[2] Petrunin A, Panfilov A, Panfilov A 2009 *Foundryproduction* **10** 17–20
[3] Reshetnikova S 2008 *Application of nanopowders of chemical compounds to improve physical and mechanical characteristics of engineering products.* Author's abstract. diss. Ph.D. Siberian State Aerospace University named after Academician M.F. Reshetnev, Moscow
[4] Kurganova Yu 2008 *Development and use of dispersed-hardened alumo-matrix composite materials in mechanical engineering.* Author's abstract. diss. Doctor of technical sciences. Institute of Metallurgy and Materials Science. A.A. Baikova, Moscow
[5] Chernysheva T, Kobeleva L, Kalashnikov I, Bolotova L 2009 Metals 1 79–87
[6] Yurchenko A et al 2015 IOP Conf. Ser.: Mater. Sci. Eng. 81 012112
[7] Lee D, Tolochko O Choi C, Kim B 2013 Powder Metallurgy 45(3) 267–270 doi: 10.1179/003258902225002532
[8] Mekhtiyev A, Zholdubaeva Zh, Issin B, Golovacheva V, Yurchenko A, Issin D 2016 METALURGIJA 55(3) 493–496
[9] Mekhtiyev A, Akhmetov A, Yudakova V, Bulatbayev F 2016 METALURGIJA 55 (1) 47–50
[10] Yurchenko A et al 2016 MATEC Web Conf. 79 01085
[11] Mekhtiyev A, Azbanbyaev E, Issagulov A, et al 2015 METALURGIJA 54(4) 623–636
[12] Shashev D V et al 2015 IOP Conf. Ser.: Mater. Sci. Eng. 81 012101

History of technical protection. 60 years in science: to the jubilee of Prof. V.F. Minin

S E Shipilov[1], V P Yakubov[2]

[1]Candidate of Physico-mathematical Sciences, associate professor the department of radiophysics, Tomsk State University, Tomsk, Russia
[2]Doctor of Physico-mathematical Sciences, professor the department of radiophysics, Tomsk State University, Tomsk, Russia
E-mail: s.shipilov@gmail.com

Abstract. The article briefly describes the scientific achievements of the full Professor, Doctor of Technical Sciences, the founder of the Institute of Applied Physics, the academician of the Russian Academy of Technological Sciences, the winner of the State Prize of the USSR Vladilen F. Minin.

1. Introduction

Prof. Vladilen F. Minin (Figure 1) was born on May 27, 1932 in the Rudinka Gorlovskogo area of the Ryazan. Dr. Minin is a graduate of the Moscow Physical-Technical Institute (MFTI). His professional career began as a teacher of night school. In 1958, at the invitation of academic M.A. Lavrent'ev, he moved from Moscow to work in the Novosibirsk Academytown.

Figure 1. Prof. Vladilen F. Minin

2. Main scientific results

His scientific activity began in Institute of Hydrodynamics where at first time he has been researching the interaction of shock waves and cavitation effect. A series of experimental investigations of the propagation of strong shock wave (SW) in a liquid with gas bubbles, which made it possible to establish the basic laws governing the process, the mechanism for the transformation of the energy of the SW, attenuation of the SW, and the formation of the structure, and experiments on the analysis of the attenuation of waves in bubble screens with different acoustic properties were begun by pioneer works of V.F. Minin at the Institute of Hydrodynamics (Novosibirsk, Russia) in the 1957th - 1960th, who examined also the first convenient model of a screen - a sequence of alternating flat one-dimensional liquid and gas layers [1].

In an experimental investigations of the dynamics of the form of pulsating gaseous cavities and interaction of SW with bubble clouds in 1957-1960 V.F. Minin [1] discovered that under the action of SW a bubble collapses asymmetrically with the formation of a cumulative jet, which forms in the process of collapse and causes fragmentation of the bubble.

Amongst studies of surface effects we should call attention to two works on the problem of modeling of an explosion on a free surface. Based on investigations of eddy dynamics Deribas et al [2] concluded that the flow arising is self-similar, and the parameter determining the effect of the explosion is the momentum imparted to the liquid. Accurate experiments, performed for the first time by Minin [3], clarified the fact that the law governing the development of the eddy, obtained in the preceding work, does not correspond to the motion of a weightless liquid. The self-similarity remains, however, and its indicator equals 0.47 and 0.38 for cylindrical and point explosions, respectively.

This problem was connected with formation of a crater and with movement of the ground in the explosion of a meteorite near the surface or upon collision with the ground. The basic mechanical properties characterizing this process are the parameters of the shock wave formed in the ground and the pattern of permanent displacements produced (in particular, the dimensions of the crater formed). The possibility of estimating shock-wave parameters for an explosion near the surface using known relationships for underground explosions has been discussed several times [3, 4].

As has been noted previously [3, 4] cavity growth in electrical explosion on the surface of water proceeds according to a law close to the law of self-similar motion at constant energy. This fact allow us to introduce the energy transferred to the ground and estimate its value (perhaps somewhat arbitrarily) by comparing the mechanical action parameters of contact and slightly submerged charges.

In the beginning of 1960th at the Institute of Hydrodynamics, Minin also laid down the experimental foundations of electromagnetic accelerating of bodies to hypersonic velocity. Since the organization of Novosibirsk State University, V.F. Minin has lectured to students of preparatory courses and organized more than 150 laboratory works on physics, some of which are still unique.

A few years later Minin investigated and discovered a wide class of jet flows in an analysis of vertical plumes on the free surface of a liquid with underwater explosions. The plumes were distinguished by the directed nature of the ejection, characteristic only for a liquid and not occurring for analogous explosions in soils.

Later in the middle of 1970th M.A. Lavrent'ev wrote "Was given that from 1958 to present Minin V.F. works at the Institute of Hydrodynamics SB USSR Academy of Sciences (since 1970 Head of the Laboratory number 3) where he are in person, with help of senior laboratory Kotov and A.I. Feoktistov, and later a senior laboratory V.M. Volosuhi, the experimental studies in the field of propagation of shock waves in water with gas bubbles, in the water with a layer of air, a study on the stability and the anomalous properties of nitric glow plasma, a study on the possibility of creating devices "Heat" and "magnet", a study on the explosive sound sources, the experimental studies on the formation of the Sultan in the explosion, other work. In other laboratories of the Institute in these areas of work not carried out. Director of the Institute of Hydrodynamics SB USSR, Academician M.A. Lavrent'ev" (Figure 2).

These effects later were studied in application to hypervelocity impact of meteorite with earth's surface [5]. Later [6] an experimental study of an explosion close to a free surface was conducted.

IOP Conf. Series: Materials Science and Engineering **363** (2018) 012033 doi:10.1088/1757-899X/363/1/012033

Principles connected with the penetration depth of the discharge were established. It was shown that the nonsymmetry of the explosion bubble compression produced by the free surface leads to the formation of vortex motion during bubble compression. Experiments were conducted under laboratory conditions using an electrical discharge and explosive charges of 1 g. Under natural conditions, charges up to 10 kg were detonated. For the plane case under laboratory conditions a fluid velocity field and particle trajectories were obtained.

Figure 2. The first Siberian seminar on explosion problem at the middle of 1970th. From left to right: Prof. V.F. Minin, Prof. L.A. Turoc and Ak. M.A. Lavren'tiev.

Based on the results mentioned above Dr. V.F. Minin's in the middle of 1970th has also presented major technological contributions in the defense against terrorism. He has created physical development principles that have resulted in the creation of a family of unique devices including explosive-protection chambers, gas detectors, x-ray control, etc. These devices were employed as part of the security system for the 1980 Olympic Games held in Moscow (Figures 3-4). Figure 4 on the right shows the first hologram there words in Russian "hologram" and "SHF".

Figure 3. Left: Explosion-protection chamber based on damping of shock waves in water with gas bubbles, middle: X-ray complex at Sheremetyevo Airport, right: X-ray inspection of a car.

Figure 4. Sound vision system (left) and the first microwave hologram (right, beginning of 1980th).

Prof. V.F. Minin has carried out in-depth studies on the creation of effective machine-sensitive methods and algorithms for numerical modeling of physics and the study of mechanics. Some of the research projects conducted by Dr. Minin include the development of automated computer programs - Program-Technical Systems (PTS) - that allows experiments to process results in the terminology of the researcher or the designer. Developed under the supervision of Prof. Minin, PTSs utilize bundles of applied programs for a variety of applications, including: non-stationary problems of mechanics of continuous environment, including explosive physics and hypervelocity impact, and systems for automated design of systems by mechanical engineering (Figure 5). The innovative special program complex is based on new high-effective method of individual particles developed under the scientific leadership of Prof. Minin. Mechanical engineering problems entailing the increase of efficiency have also been solved with the assistance of high-speed computing processes developed by Prof. Minin. In addition the ability to fundamental model protection of the "Vega" space vehicle from potential meteoric hypervelocity impacts up to speed of 30 km/s with wide-range equation of state of target was developed (Figure 6). The method of computation, calculation experiment technologies and some of examples of application in explosive and hypervelocity physics are described at [7–11].

In 1988, for his work in the creation of computer facilities and technological advancements in computer experiments, Prof. V.F. Minin was awarded The State Medal of the U.S.S.R., 1st Class and he was the winner of the Statepremium of the USSR for the developed and investigation of the hypervelocity meteorite shield concept for the "Vega" spacecraft.

Prof. Minin investigated Hypervelocity impact phenomenons also in application to the problem of thermonuclear fusion and compressing of different material under high-pressure conditions [12–13].

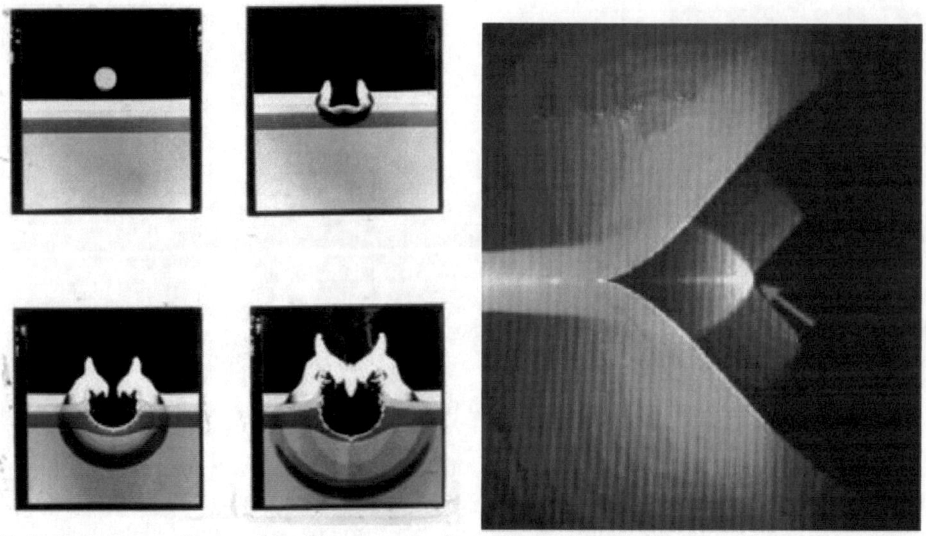

Figure 5. Left: Impact of asteroid with ocean surface, Right: Mach wave generator (the right picture was also used in: V. Fortov. Intense shock waves and extreme states of matter. International Albert Einstein Gold Medal Lecture, but without specifying the authors of the simulation).

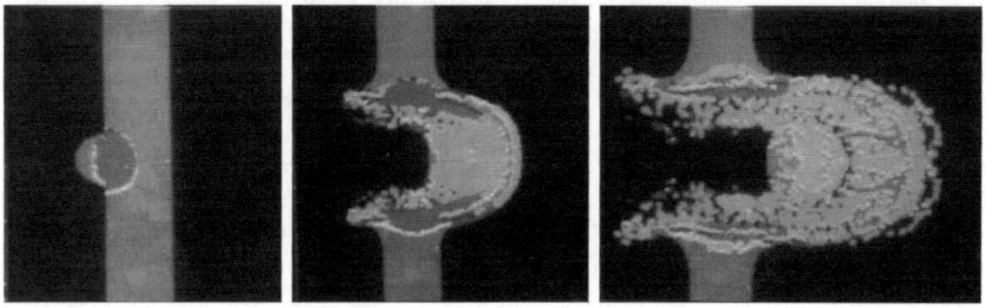

Figure 6. Simulation of the macroparticle impact (V~30 km/s) with aluminum screen: blue – solid, red – liquid, green – plasma states of matter.

Fundamental studies of shaped charge hypervelocity jet formation allow formulating by Prof. Minin the following main conditions: it was shown, that in axisymmetrical case the criterion of Walsh of jet formation is inapplicable; In axisymmetrical case the cumulative jets are formed always [14].
These results allow to formulate an innovative concept of so-called "forced jet formation" for hypervelocity shaped charge jets which has a big advantages before classical shaped charge jets (big mass of jet, speed, the jet without slag, hypervelocity interaction with different targets and so on) [7, 15, 16]. Based on these principles Prof. Minin developed innovative fundamental physical principles of hypercumulative shaped charge jet formation which haven't analogies in the World for today [7, 15, 17]. These principles allow increasing both as the speed as a mass of hypervelocities jets in several times.

The fundamental physical effects accompanying the flight of cumulative jets at hypersonic speeds in the air and air penetration were studied. The influences of a grain size of microstructure of the surface layer material of a hypersonic body on the properties of air plasma were studied at the first time under the scientific leadership of Prof. Minin. Some new physical effect were discovered when hypervelocity body forming by shaped charge moving in the air with normal pressure (the effects of erosion from the surface of the jet, but not an ablation) [18].
Some of Prof. V.F. Minin many citations and achievements include; The Red Banner Award of Labor, he is the winner of the State premium of the USSR, academician of the Academy of Technical Science

of The Russian Federation, he is the founder and President of the Uralo-Siberian branch of the Academy of Technological Sciences of Russia. He is the winner of the Order of Lenin as well as numerous other awards and medals for his scientific and academic achievements. Dr. Minin has also authored the openings of more than 600 technical proceedings, several monographs and holds patents on more than 150 inventions.

3. Conclusion

In conclusion, we would like to wish Prof. Vladilen F. Minin Siberian health, talented students for further research and getting new discovers and scientific results!

References

[1] Minin V F 1959 The interaction of underwater shock wave with a bubble curtain. PhD dissertation (Novosibirsk) p 117
[2] Deribas A A and Pokhozhaev S I 1962 *Dokl. Akad. Nauk SSSR* **144**
[3] Minin V F 1964 *Zh. Prikl. Mekh. Tekh. Fiz.* **3**
[4] Dienes J K and Walsh J M 1970 *Academic Press*
[5] Minin V F, Alekseev A S, Kryukov B P 1989 The cosmogenios factors of effect on an earth's surface *Scientific seminar on mathematical modelling and prediction natural and technogenios catastrophes* Computer centre SB AS USSR
[6] Minin V F and Kuznetsov V M 1970 *Proc. of the Seminar of the Institute of hydrodynamics* Siberian Branch of Academy of sciences of the USSR October 21
[7] Minin V F, Minin I V and Minin O V Calculation Experiment Technology. In: Computational Fluid Dynamics Technologies and Applications, Eds. by Igor Minin and Oleg Minin. IN-TECH, Austria 2011, available at: http://www.intechopen.com/articles/show/title/calculation-experiment-technology
[8] Minin V F, Minin I V and Minin O V 1992 The calculation experiment technology *2 Int.Symp.on intence Dynamics loading and its effects Chengdu*, China Jule 9–12
[9] Minin V F 1989 *Preprint of the Institute of control problems* 1–54
[10] Agureikin S I, Anisimov A V, Bushman G I, Kanel V P, Karyagin A B, Konstantinov B P, Kryukov V F, Minin S V, Razorenov R Z, Sagdeev S G, Sugak V E Fortov 1984 *High Temperature* **22** 761–778
[11] Minin V F, Alekseev A S, Kryukov B P 1989 The cosmogenios factors of effect on an earth's surface. - *Scientific seminar on mathematical modelling and prediction natural and technogenios catastrophes* (September 15-16) Computer centres SB AS USSR
[12] Minin V F *Letters to Russian journal of technical physics* **14** 1765–1769
[13] Minin V F Hydrodynamic effects in conic thermonuc-lear targets 1987 *18-th ECLIM* Prague (May 4-8) 187
[14] Minin V F, Minin I V and Minin O V 1992 Analytical and computation experiments on forced plasma jet formation. *Proc. of the Int. Symp. on Intense Dynamic Loading and Its Effects.-* Chengdu, China 588-591
[15] Minin V F, Minin I V and Minin O V 2013 *International Journal of Modern Applied Physics* **2** 130–141
[16] Minin V F, Minin I V and Minin O V Method and devices to forming hypercumulative jets for oil well perforation with big holes and without slug. Patent of Russian Federation 241338 from 7.12.2009)
[17] Minin V F, Minin I V and Minin O V 2012 *11TH Int. Conf. on actual problems of electronic Instr. Eng., (APEIE)*
[18] Minin I V and Minin O V 2000 *The 10th Electromagnetic Launch Technology Symposium* California USA

Web technologies for rapid assessment of pollution of the atmosphere of the industrial city

N Shaparev[1,2], A Tokarev[1], O Yakubailik[1,3] , A Soldatov[2]

[1]Institute of Computational Modeling SB RAS, Krasnoyarsk, Russia
[2]National Research Tomsk State University, Tomsk, Russia
[3]Siberian Federal University, Krasnoyarsk, Russia
E-mail: oleg@icm.krasn.ru

Abstract. The functionality, architectural features, the user interface of the geoinformation web-system of environmental monitoring of Krasnoyarsk is discussed. This system is created in service-oriented architecture. Data collection from the automated stations to monitor the state of atmospheric air has been implemented. An original device to measure the level of contamination of the atmosphere by fine dust PM2.5 has developed. Assessment of the level of air pollution is based on the quality index AQI atmosphere.

1 Introduction

The quality of atmospheric air is the most important environmental factor determining the health of population and the state of ecosystems. Air monitoring systems have been created nowadays in almost all large population centers and industrial zones in Russia and other countries of the world; they control dozens of indicators and concentrations of various chemicals (pollutants) in the atmosphere.

Taking into account the multi-pollutant nature of the atmospheric air contamination and its dynamism and spatial heterogeneity, an important task is to characterize in a comparable and clear form the level of total (integral) air pollution averaged over different periods of time and the health risks of population of a certain territory [1–4].

This scientific work examines the experience of work related to monitoring of the atmospheric air pollution in Krasnoyarsk, an industrial city with population exceeding 1 million people. Krasnoyarsk is one of the most polluted regions in the country according to reports of Federal Service for Hydrometeorology and Environmental Monitoring of Russia. The city monitoring is carried out by a number of organizations at the federal and regional levels (the Federal State Budget Office "Central Siberian Department for Hydrometeorology and Environmental Monitoring", the Regional State Budget Office "Center for Implementation of Environmental Management and Environmental Protection of the Krasnoyarsk Territory", the Main Directorate of the Ministry of Emergency Situations of the Krasnoyarsk Territory, Siberian Federal University, research institutes of the Russian Academy of Sciences, etc.). Each of these organizations has its own methods, technologies, and systems for data collecting, storing, and processing. In addition, with appearance of the available instruments for assessing the air pollution, public environmental organizations and independent activists and bloggers have become involved in collecting information on pollution levels. The variety of the applied solutions, interdepartmental and organizational disconnection lead to the fact that a complex analysis and operative evaluation of the entire array of the recorded information is currently technically difficult and practically not carried out. Taking into account the expanding number of stationary posts that monitor the state of the natural environment in the coming years, planned by the authorized organizations, the situation will only worsen [5, 6].

Research and development for the atmospheric pollution monitoring carried out at the Institute of Computational Modeling of the Siberian Branch of the Russian Academy of Sciences (ICM SB RAS) are aimed at solving this problem. At first, a web-based geoinformation analytical system for monitoring the atmospheric pollution is being developed. Its distinctive feature is presentation of information about the level of atmospheric pollution in a simple visual form, based on the atmosphere quality index (AQI).

Secondly, it solves the problem of collecting information on the level of urban atmospheric pollution, this information comes from different sources into a centralized database. Within the framework of the organized information interaction with the authorized regional subordinate organizations, the information about the state of the atmosphere is collected in real time from the automated monitoring posts. A web-based interface has been developed for analytical processing and data presentation.

Third, based on the available components, a working prototype of a device for measuring the level of atmospheric pollution by the fine dust (particulate matter) has been developed, and its small-scale production is being prepared. The created device provides automatic transmission of contamination level data into the monitoring system database via the cellular network.

2 Air Quality Index

For this integrated assessment, in many countries around the world, complex indicators, i.e. various types of the air pollution index, have been developed and put into practice nowadays [7]. The main areas of application of such indexes are the integrated assessment of the level of the air pollution in the city as a whole or its areas, the comparison of the levels of the air pollution, identification of the long-term changes in the quality of the atmospheric air. These indexes are used as a tool for decision-making in environmental management [8–10].

In particular, the Canadian Meteorological Service uses the AQI air quality index, for which the scale of values varies from 0 to 100, for presentation of the air monitoring results. The higher the index value, the higher the risk to human health. The index is calculated for six key pollutants: SO_2, O_3, NO_2, the total content of the reduced sulfur compounds, CO and suspended particles (PM – particulate matter). Using the monitoring stations network data, the Ministry of the Environmental Protection determines the AQI for each territory on which monitoring stations are located. These information is communicated to the public and media on a daily basis in a timely manner. The AQI scale is as follows: 0–25 is a good quality (green color); 26-50 is moderate (yellow); 51–100 is poor (orange); and over 100 is very poor quality (red color).

In France, the ATMO index is used for integrated assessment of the atmospheric air pollution. It is regulated at the national level by a special Decree of the Ministry of Ecology and Environmental Protection. The ATMO is calculated by the contents of four pollutants: SO_2, NO_2, O_3, and particulate matter 10 microns and below (PM_{10}). For each pollutant, the primary index is calculated, and the daily atmospheric pollution index is calculated by summing the highest primary indexes.

The UK Meteorological Department publishes the air quality forecasts, where the level of the air pollution is described by index (from 1 to 10) and the corresponding pollution level (from 1 to 3 – low, from 4 to 6 – moderate, from 7 to 9 – high, 10 – very high). These levels are established on the basis of how each pollutant affects people's health. During the index calculation, the concentrations of SO_2, NO_2, O_3, and PM measured at the stations are taken into account.

In Hong Kong since 1995, the API Index is used to assess the air quality. Two API types are calculated: aggregate (General) and applicable for transport roads (Roadside). The Department of Environmental Protection evaluates hourly both types of the API indexes and builds the API forecast for the next day. At the same time, it is calculated on the basis of levels of six contaminants: SO_2, NO_2, suspended particles PM, CO, O_3, Pb, measured at the monitoring posts. At values of API up to

25, the level of atmospheric air pollution is considered low, 26–50 is medium, 51–100 is high, 100-500 is very high, and 201–500 is considered extremely high.

The Environmental Protection Agency (EPA) of the United States and its regional offices regularly calculate and publish the Air Quality Index (AQI). AQI is a tool for providing information about the air pollution to the wide public in a simple and visual form. The basic idea is that for each substance a scale of pollution levels is formed, consisting of several classes, depending on the degree of the impact on human health. AQI calculation is based on concentration indexes of several pollutants, i.e. suspended particles of less than 10 μm in diameter (PM_{10}) and less than 2.5 μm ($PM_{2.5}$), carbon dioxide CO, gas SO_2, NO_2, and ozone O_3. For each class of the pollution level scale, a color designation is also introduced (green / yellow / red / burgundy / black means the corresponding degree of the air pollution and health effects, ranging from a safe level to a natural disaster), followed by recommendations to the public.

The China Environmental Monitoring Center used a variety of integrated pollution assessment methods for a number of years, but currently presents data in the form of AQI in a way similar to the US EPA; the differences are mainly in the choice of the averaging period for the air pollution data.

It should also be mentioned, that in the US, China and other countries of the world, along with the AQI index, auxiliary indexes, such as InstantAQI, NowCast, NowCastChina, and others are used. The NowCast is a weighted average of hourly air monitoring data used by the United States Environmental Protection Agency for real-time reporting of the Air Quality Index (AQI) for PM (PM10 or PM2.5) or Ozone data. The PM NowCast is computed from the most recent 12 hours of PM monitoring data, but the NowCast weights the most recent hours of data more heavily than an ordinary 12-hour average when pollutant levels are changing. The PM NowCast is used in lieu of a 24-hour average PM concentration in the calculation of the AQI until an entire calendar day of hourly concentrations has been monitored [11,12].

3 The system of monitoring the data collection, presentation, and analysis

The source of the data about the level of the atmospheric pollution is the Regional Departmental Information and Analytical System on the State of the Environment of the Krasnoyarsk Territory (RIAS), which has been collected since 2009. The system operator is the Regional State Budgetary Institution "Center for Implementation of Environmental Management and Environmental Protection Measures in the Krasnoyarsk Territory". Observations on the quality of the atmospheric air are carried out at 8 automated observation posts, 6 of which are located in Krasnoyarsk. These posts are equipped to provide continuous automatic measurement of mass concentrations of oxide and nitrogen dioxide, sulfur dioxide, carbon monoxide, dust, formaldehyde in the ambient air, as well as collection, processing, storage, and transfer of the accumulated information to a remote server. Along with measurements of the atmospheric concentrations of pollutants, the meteorological parameters are measured in automatic mode at meteorological posts (wind direction and speed, temperature, humidity, atmospheric pressure). In addition, in some Krasnoyarsk areas monitoring of the air pollution, with application of the mobile laboratory on a sliding schedule, for detecting pollutants of ammonia, hydrogen sulfide, hydrochloride, hydrofluoride, benz(a)pyrene and suspended substances are daily conducted [13].

Implementation of the data collection system is based on the software tools of the ICM SB RAS geoportal. Access to observational data is carried out by means of standard geoportal tools, including viewing of the tabular data, exporting, viewing of the data on maps with possibility to select time intervals and access with application of generally accepted standards. A new resource type i.e. "observational data" was added to the geoportal resource catalog. When publishing this resource on the geoportal, the user is provided with a number of settings, including selection from the list of the available monitoring stations and sensors. Each observation post has a spatial reference and a certain set of sensors. In this case, different observation posts, not belonging to the same group, may have several common sensors [14, 15].

To view the observation data, the capabilities of the existing cartographic web interface of the geoportal have been extended. The key elements of the new interface include a background base map and observation data in the form of a semitransparent layer at the selected time point for the selected indicator. The user of the web application is provided with such control elements as the choice of one of the indicators and the time interval. With the help of the additional tools, the data with a certain time step in one direction can be viewed. To search the anomalies, the data output is provided in the form of an active graph of the maximum values with a rapid transition to the viewing data at a certain point in time.

To download the data to the geoportal, a software module with the appropriate "driver" for processing and transforming the input data is prepared; it provides the observational data periodic loading via the web service. The web service provides access to the data in json format and contains three sections of information: observation posts and their coordinates, a list of indicators (pollutants and meteodata), the value of indicators with reference to the time. To minimize the load on the remote server at loading of the archive of observations from 2009 to 2017, the text files in json format are generated. The data import directly from the database on a remote server to the geoportal server took several days. After the archive import, the data is loaded once per hour. The information from the mobile laboratories is downloaded every few days. When import problems arise from a remote web service, the developed software automatically finds the date of the last received data and forms a request for new data download. To minimize the load on the remote service, the data is loaded in portions at a certain time interval until the current data time is reached.

4 Instrument for measuring the air pollution level

Design and development of a device for recording the concentration of finely dispersed suspended particles up to 2.5 μm in size ($PM_{2.5}$) in the air was made on the basis of the available electronic components on the market. The system is based on the Arduino microcontroller, an electronic designer and a convenient platform for rapid device development, which is very popular all over the world owing to convenience and simplicity of the programming language, as well as open architecture and software code. The microcontroller is programmed with the use of the Wiring language (a simplified version of C ++) and the Arduino IDE development environment.

Arduino Nano 3.0 board was chosen from a number of different microcontroller modifications, based on sufficient memory, I/O ports, miniaturized size and convenience. The following components were also used: the GSM SIM800L cellular communication module for data transmission to the geoportal, the data modules via Ethernet and Wi-Fi, the OLED SSD1306 display for the current values, the AMS1117 linear regulator at 3.3 V, LogicLevelConverter, power supply.

The key element of the device is the PMS7003 module, a universal digital sensor, developed by the Plantower company for measuring the concentration of suspended air particles [16]. The basis of this sensor is measurement of the laser radiation scattering on suspended particles. Using the built-in microcontroller, based on the physical model of scattering of a spherical, homogeneous, isotropic, and non-magnetic particle in a nonabsorbing medium, the equivalent particle diameter and the number of particles with different diameters per unit volume are calculated. The sensor has a digital serial interface, using which the measured values of dust concentrations are fed to the Arduino microcontroller.

The developed device can function in 2 modes, passive and active. In the active mode, the sensor periodically outputs the measurement data. It is divided into two sub-modes: stable and fast. If the concentration of suspended particles is small, the sensor will operate in stable mode with a real update interval of 2.3 seconds. Otherwise, the sensor will operate in a fast mode with an update interval of 200-800 ms (the higher is the concentration – the shorter is the interval). In passive mode, the data is issued only after a preliminary request. The device requires an external power supply, connected via a standard micro-USB connector (Figure 1).

IOP Conf. Series: Materials Science and Engineering **363** (2018) 012034 doi:10.1088/1757-899X/363/1/012034

Figure 1. Prototype of the device for measuring the level of the atmospheric pollution by particulate matter.

5 Atmospheric pollution monitoring system

Design and development of the web GIS monitoring user interface was carried out on the basis of the application program interfaces (API) of the ICM SB RAS geoportal, a set of previously developed software tools. To design the web application's screen forms, a mechanism for generating dynamic content based on templates was used. Templates allow to change the order and the form of the attribute data output according to the objects in map layers, including different elements of style design, i.e. color, font parameters, etc. The developed monitoring system is currently in trial operation [17, 18].

To collect and process the data on the air pollution by suspended particles, the SensorCollector service of the ICM SB RAS geoportal is used. The service API is built based on the REST approach, and the requests are transmitted by HTTP methods GET / POST / DELETE with parameters. The transmission of the current values of the indicators is performed by the following query:

http://gis.krasn.ru/sc/api/1.0/projects/<project_id>/values/send?key=<key>&
site=<site_id>&<values>, where:

- key is the user's key to access the service;
- project_id is the project identifier within which the data are collected;
- site_id is the identifier of the site on which the sensor is located;
- values are values of indicators transmitted by separate parameters in the following form: <code> = <value> & ... <code> = <value>, where code is the indicator code in the system, and value is its numerical value.

6 Preliminary monitoring results

The developed software and hardware tools to support the archives of observations of the environment state in Krasnoyarsk provided the opportunity to conduct a spatio-temporal analysis of the ecological state of the city.

Krasnoyarsk is the most eastern Russian city with population exceeding million people. The climatic conditions are very unfavorable for dispersion of impurities. The city is located in the zone of high potential of atmospheric pollution (PAP). Frequent air stagnations lead to accumulation of impurities in the atmosphere and formation of high levels of the air pollution. Presence of the aluminum plant, three coal-fired power plants and a large number of vehicles contribute to the atmosphere pollution. In particular, in 2015 the total emissions amounted to 195 thousand tons, including 128.7 thousand tons from stationary sources [19].

The city residents call the visually visible aerosol pollution of the atmosphere residents as "black sky", although the correct name of this phenomenon is could. Monitoring of the PM_{10} concentration in the surface layer of the Krasnoyarsk atmosphere has been made since the end of 2012.

Services of the geoportal operational monitoring subsystem data provide tools for data analyzing and presenting. In particular, based on the available data, a preliminary conclusion can be made, that concentration of suspended particles (PM_{10}) in the surface layer of the atmosphere at 1.10–1.68

exceeds the average annual maximum permissible concentration (MAC). The analysis shows that in Krasnoyarsk there is clearly lack of the automated observation stations (AOS) that could provide a reliable estimate of the level of aerosol pollution of the atmospheric air in a metropolis. In those city areas where the AOSs are located, the PM_{10} concentration in most cases exceeds the average annual MAC. It can be interpreted as aerosol pollution in a certain region of Krasnoyarsk. In 2015–2016, the aerosol pollution was recorded at suburban AOS as well.

With calm and weak wind, the aerosol pollution is observed over the territory of almost entire city. At the same time, all AOSs record PM_{10} concentrations exceeding the maximum single permissible concentration. Similar situations are realized in the unfavorable meteorological conditions period, which in Krasnoyarsk reach 60 days per year.

Figure 2 shows the values of daily average and maximum single concentrations of suspended particles (PM_{10}) in 2016. Apparent is the excess over the maximum permissible concentrations; for PM_{10} they are as follows: maximum single = 0.3 mg/m^3, and daily average = 0.06 mg/m^3.

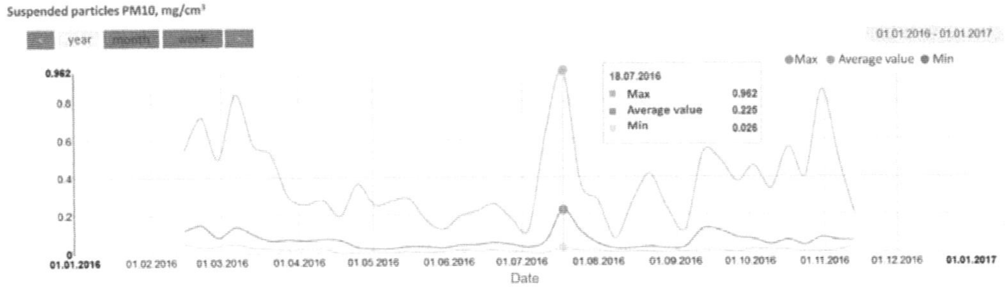

Figure 2. Dynamics of concentration of suspended particles of 10 μm (PM10) in 2016.

7 Conclusion

Work has been done on design and development of information and analytical software for monitoring and assessing the quality of the atmospheric air. The software components for calculating the AQI atmosphere quality index are developed; they are integrated into the subsystem of operational and scientific research monitoring over the state of pollution of the ICM SB RAS geoportal. Application web services have been created to automatically download the data from the edge system of observations of the state of the atmospheric pollution into the geoportal database. A web application has been developed to visualize this data.

A model of a device, based on Plantower PMS7003 sensor and Arduino Nano microcontroller, for measuring PM2.5 suspended particles concentration in the atmosphere, was created. The device logics and control modes are implemented with the help of firmware in the Wiring (Simplified C ++) language with application of a number of open source libraries. This device provides reflection of the current values of contamination levels on the OLED display and transfer of the recorded data to the geoportal through the cellular network. The geoportal implements a specialized web service for receiving the data and placing them into the operational monitoring database.

The studies are conducted, and results of the spatial distribution of the atmospheric pollution in Krasnoyarsk are obtained according to the official environmental reporting. As a result, a geospatial data base was formed, on the basis of which the integrated maps of emission sources for city pollutants were constructed, taking into account, among other things, the terrain inhomogeneity.

References

[1] Qilu Li, Kong Yang, Kechang Li, Xin Liu, Duohong Chen, Jun Li, Gan Zhang 2017 *Environmental Pollution* **224** 679–688

[2] Sharma M, Maheshwari M, Sengupta B, Shukla B P 2003 *Environmental Modelling and Software* **18** 405—411
[3] Brian R P, Daniel J, Mehaffey M, Jackson L E, Neale A 2015 *Ecosystem Services* **14** 45–55 doi: 10.1016/j.ecoser.2015.04.005
[4] Lanzafame R, Monforte P, Patanè G, Strano S 2015 *Energy Procedia* **82** 708–715
[5] Kadochnikov A A, Yakubailik O E 2015 *NSU Journal of Information Technologies* **1** 37
[6] Yakubailik O E, Popo V G 2009 *Computational Technologies* **6** 116–126
[7] McCarty J, Kaza N 2015 *Landscape and Urban Planning* **139** 168–179
[8] Gayathree Devi P K, Sujatha C H 2016 *Atmospheric Pollution Research* **7** 1053–1064
[9] Wu J, Cheng J, Zhang L, Chen M, Pan L, Chen X *Journal of Computational Information Systems* **10** 9839–9848
[10] Ayushi Vyas, Siby John 2016 *MATEC Web of Conferences* **57** 05002
[11] Rabee A M 2015 *Environmental Monitoring and Assessment* **187** 4203
[12] Ling Xue, Yeping Zhu, Yan Xue, 2013 *Mathematical and Computer Modelling* **58** 480
[13] Shaparev N, Yakubailik O 2016 MATEC Web of Conferences **79**, 01081
[14] Kadochnikov A A, Popov V G, Tokarev A A, Yakubailik O E 2008 *Journal of Siberian Federal University. Engineering & Technologies* **4**, pp. 377–386
[15] Yakubailik O, Kadochnikov A, Tokarev A 2015 *Scientific GeoConference SGEM* **1** 487
[16] Korose C P, Locke Ii R A, Blakley C S, Carman C H 2014 *Energy Procedia* **63** 3945–3955
[17] Mekhtiyev A *et al* 2016 METALURGIJA **55** 47–50
[18] Yakubailik O E 2010 *Vestnik of SibGAU* **1** 40–45
[19] Yakubailik O E, Gosteva A A, Erunova M G, Kadochnikov A A, Matveev A G, Pyataev A S, Tokarev A V 2012 *Vestnik of KemSU* **52** 136–142

Video control system for a drilling in furniture workpiece

V L Khmelev, R N Satarov, K V Zavyalova
Tomsk State University, Tomsk, Russia
E-mail: XmeJleB@mail.ru

Abstract. During last 5 years, Russian industry has being starting to be a robotic, therefore scientific groups got new tasks. One of new tasks is machine vision systems, which should solve problem of automatic quality control. This type of systems has a cost of several thousand dollars each. The price is impossible for regional small business. In this article, we describe principle and algorithm of cheap video control system, which one uses web-cameras and notebook or desktop computer as a computing unit.

1. Introduction

Machine vision system consists of defined tasks such as counting objects on a conveyor, reading serial numbers, and searching for surface defects. Manufacturers often prefer machine vision systems for visual inspections that require high speed, high magnification, and repeatability of measurements. The system using in a different part of industry for solving different type tasks. In our case, we had tried to solve problem of control drilling operations. A control of drilling operations is set of tasks: location and size control of holes, the shape and size of the workpiece. In other hand, we understand that most of systems (from famous companies as “Datalogic” or “Cognex”) direct to high price market. It pushed us to try use cheap webcamera (with a price about 20 US dollars) as base element of system. Such step make a possible using this type of system for small business and/or start-ups.

Nevertheless, using cheap camera make us calibrate it. Geometry calibration estimates the parameters of lens and image sensor of video camera. We have to use these parameters to correct lens distortion, and high precision measure the size of an object in world units [1].

Our system consist of web camera fixed to a rod, the camera is strictly oriented to a measure table, shown on Figure 1.

Figure 1. The measure table with fixed web-camera.

An operator places finished worpiece on the table. Next, the camera catches picture of the workpiece and processing it, comparing with original scheme. After processing, an operator will get a grayscale image of workpiece, where he sees color legend of analysis, shown in the Figure 2.

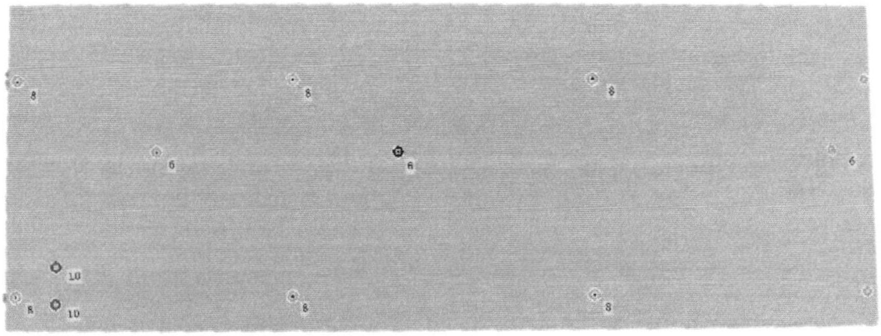

Figure 2. Grayscale image of workpeice with color legend.

Legend has a color graduation. Where green halo is successful matching between drawing and workpiece for location and diameter of holes. Red halo is holes, which exists in the origin scheme, but skip in the workpiece. The last, blue halo is opposite to red; hole exists in the workpiece, but skip in the origin scheme. Moreover, if diameters not matching, then the system uses red halo and show the difference of diameters.

2. Algorithm

Let's descry algorithm of processing, shown in the Figure 3. The first stage is the recognition of the drawing. We work with SVG (Scale Varialble Graphic) format drawings, as one of the most common and simple for text concertation. From drawing, we get three parameters: the shape of workpiece, the radius of holes, and the location of hole centers. At the same time, system is processing workpiece snapshot. Image processing consists of the following steps. Subtraction of the background (measuring table) [2]. Recognition of a workpiece contour [3].

Further, inside the contour, system recognize the holes in a given range of radii [4]. After it, system matching data from origin scheme and image processing. The last block take decisions for each hole.

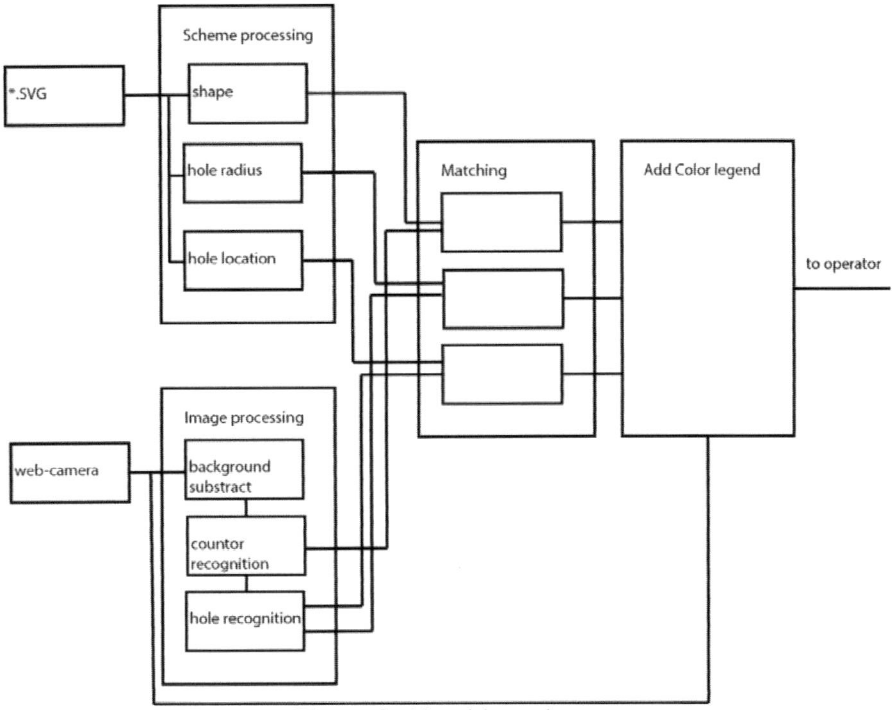

Figure 3. Algorithm scheme.

3. Conclusion

Thus, we developed the prototype and the algorithm for video controlling of drilling operations, where cheap web camera (cost about 20$) was used as a recording device. The computational block was a general-purpose computer. Computations were made in the software package MATLAB Image Processing. The introduction of the system to the furniture industry will reduce the level of losses due to drilling defects. In addition, we plan to use this system in anther projects for automatic process control. For example, will try to improve radio-tomography research [5] using this prototype as aim system.

By this article, the authors would like to emphasize a necessity of R&D support for small and medium-business automatization in technical department of Universities.

Acknowledgments

This research was supported by “The Tomsk State University competitiveness improvement programme” grant.

References

[1] Heikkila J, and Silven O 1997 *IEEE International Conference on Computer Vision and Pattern Recognition*
[2] Zhao Cong *et al* 2011 *EURASIP Journal on Image and Video Processing* **2011**
[3] Gonzalez R C *et al* 2004 *Digital Image Processing Using MATLAB*
[4] Atherton T J, Kerbyson D J 1999 *Image and Vision Computing* **17** 795–803
[5] Shipilov S E *et al* 2017 *Optical and Quantum Electronics* **49** 339